INTERPRETING CHEST X-RAYS

Illustrated with 100 cases

Philip Eng

and

Foong-Koon Cheah

Singapore General Hospital

CAMBRIDGE
UNIVERSITY PRESS

CAMBRIDGE UNIVERSITY PRESS
Cambridge, New York, Melbourne, Madrid, Cape Town, Singapore, São Paulo

Cambridge University Press
The Edinburgh Building, Cambridge CB2 2RU, UK

www.cambridge.org
Information on this title: www.cambridge.org/9780521607329

P. Eng and F.-K. Cheah 2005

First published 2005
Reprinted 2005

Printed in the United Kingdom at the University Press, Cambridge

A catalog record for this book is available from the British Library

ISBN-13 978-0-521-60732-9 paperback
ISBN-10 0-521-60732-9 paperback

CONTENTS

PREFACE

This book arose because of the huge amounts of clinical material that pass through the Singapore General Hospital, the largest tertiary care hospital in Singapore. A significant proportion of our patients come to us for a second opinion from the neighboring countries. Often they come to consult us for an abnormality on a chest radiograph. Pulmonary Medicine is largely based on the strong foundation of the plain chest radiograph. Indeed, chest radiography is the single most common investigation carried out in hospital practice. This book is targeted towards final-year medical students and residents in a medical training program. We have given countless tutorials to generations of medical students, residents, and fellows and we hope that this collection of pearls can help make the learning process easier, more enjoyable, and less painful.

Readers are advised to read this book from cover to cover as the cases are laid out in an increasing order of complexity. The latter cases assume some fundamental knowledge which is laid out in the earlier cases. The authors have intentionally made the cases as clinically relevant as possible so that interest is sustained and the book will not be heavy going.

P. ENG
F.K. CHEAH

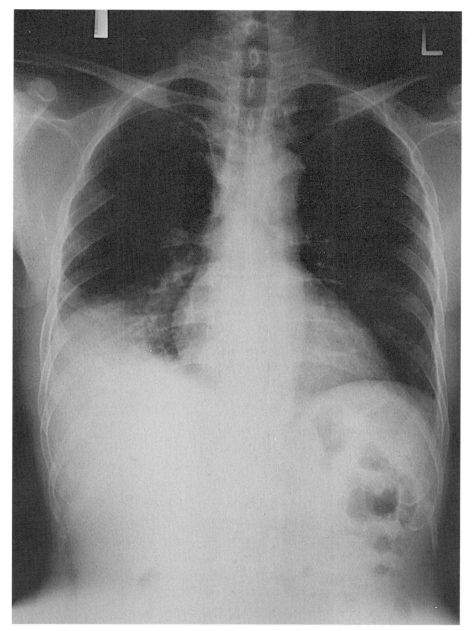

Fig. 1.1

Case 1. A 35-year-old male presented with fever, cough, and purulent sputum for one week. This was his CXR (Fig. 1.1). What is the diagnosis?

Fig. 1.2

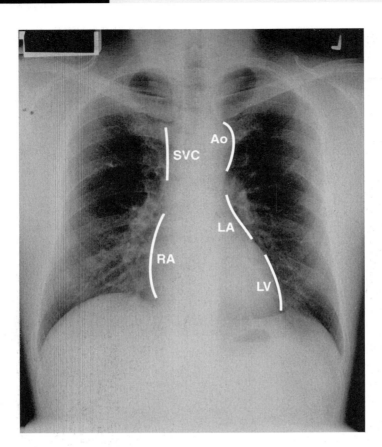

CASE 1 PNEUMONIA

The CXR shows a focal shadow in the right lower lobe with air bronchograms suggestive of pneumonia. It is clearly in the right lower lobe because the right hemidiaphragm is effaced. Right middle lobe shadows would efface the right heart border.
The presence of air bronchograms indicates pathology in the alveoli, as the con-
ducting airways remain patent with air. Water or blood can also occupy the alveoli
as a result of pulmonary edema or pulmonary hemorrhage respectively. There
should be other supporting signs such as cardiomegaly, upper lobe diversion, and
Kerley B lines with pulmonary edema. The differential diagnoses of a focal shadow
with air bronchograms include bronchoalveolar cell carcinoma and lymphoma. It
is important to follow-up the CXR to ensure that total resolution of infection
occurs. This may take up to three months in the elderly but generally some
improvement usually occurs within a week. The borders of the heart on a PA CXR
are shown in Fig. 1.2. SVC – superior vena cava, RA – right atrium, Ao – aortic
knuckle, LA – left atrium, LV – left ventricle

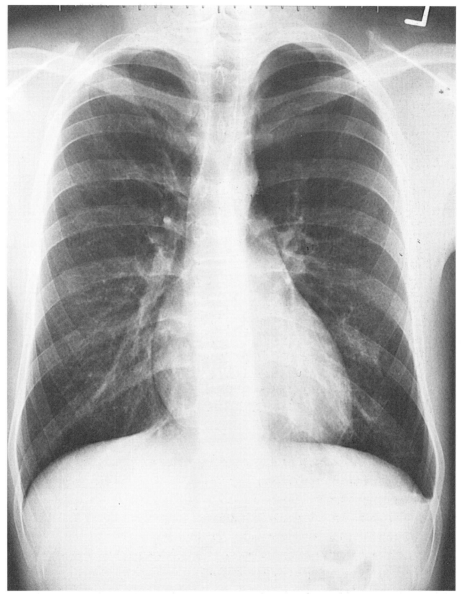

Fig. 2.1

Case 2. This 25-year-old had sudden onset of left-sided chest pain. The CXR is shown (Fig. 2.1).

Fig. 2.2

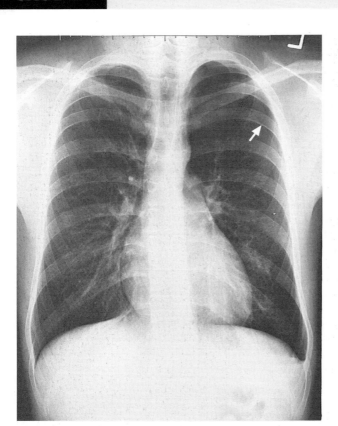

CASE 2 LEFT PRIMARY SPONTANEOUS PNEUMOTHORAX

The CXR shows the visceral pleura (Fig. 2.2) separated from the parietal pleura by air which now occupies the potential space in the pleural cavity. The visceral pleura must not be mistaken for skin-fold shadows which usually occur in supine or obese patient CXR. In addition, the line from skin folds can be seen to cross the chest wall. In the patient above, the lungs appear otherwise healthy and this condition is called primary spontaneous pneumothorax. It occurs classically in young males. This is in contradistinction to secondary pneumothorax which occurs in diseased lungs, e.g. chronic obstructive pulmonary diseases (COPD). Pneumothorax in an erect film is usually seen at the apex. See Case 60.

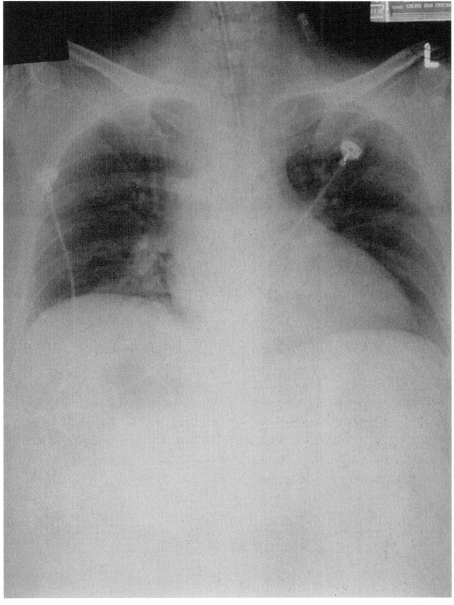

Fig. 3.1

Case 3. 50-year-old male presented to the Emergency Room with shock and a four-day history of a febrile illness. He required intubation and was started on inotropes. This was his CXR (Fig. 3.1).

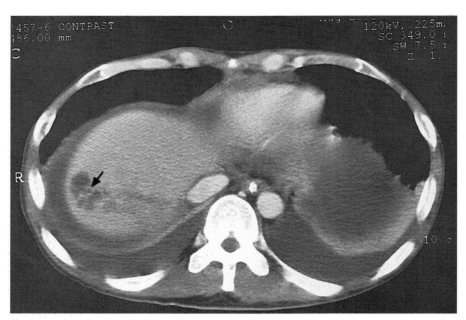

Fig. 3.2

CASE 3 RUPTURED LIVER ABSCESS

It is important to look at the "blind areas" of the CXR in order not to miss important clues. These areas are under the diaphragm, behind the heart, the hilum, and the soft tissues. This CXR shows a lucency over the liver density. The lucency does not conform to the usual bowel configuration. In this clinical context, an important differential diagnosis to be considered is a ruptured liver abscess. This can be confirmed either by bedside ultrasound or CT (Fig. 3.2). Liver abscesses are usually due to organisms like *Klebsiella* or *Amoebiasis*. All patients with *Klebsiella* bacteremia of unknown origin should have imaging studies of the abdomen to rule out a liver abscess.

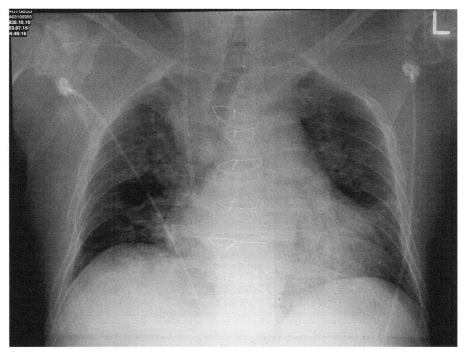

Fig. 4.1

Case 4. This elderly male has exertional dyspnea, orthopnea, and parox-ysmal nocturnal dyspnea. His CXR is shown (Fig. 4.1).

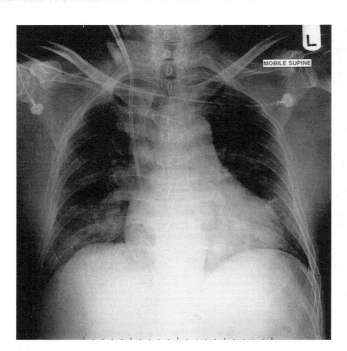

Fig. 4.2

CASE 4 CONGESTIVE HEART FAILURE

The CXR shows classic evidence of left ventricular failure, i.e. cardiomegaly (cardiothoracic ratio >50%), upper lobe pulmonary venous diversion, and Kerley B lines (which indicate distension of lymphatics). In addition, there is evidence of sternotomy wires, suggesting previous coronary artery bypass surgery (CABG). Following diuresis, the pulmonary infiltrates have cleared (Fig. 4.2). Only fluid and blood on the chest radiograph can clear rapidly (within days). This patient also has a right internal jugular central venous line.

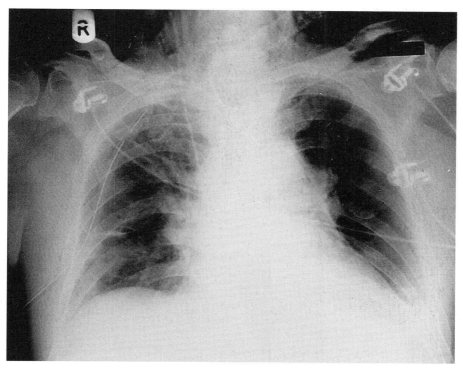

Fig. 5.1

Case 5. A 65-year-old male presented with cardiogenic shock. He had an emergency CABG which was associated with a very stormy peri-operative period. This was his CXR (Fig. 5.1) taken upon arrival at the Intensive Care Unit (ICU). What is the most significant abnormality?

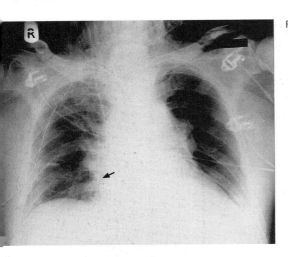

Fig. 5.2

Fig. 5.3

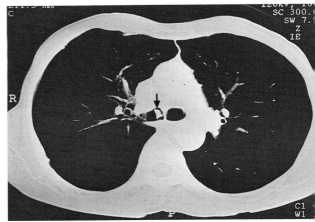

CASE 5 FOREIGN BODY RIGHT LOWER ZONE

The CXR shows an opaque density in the region of the right lower zone (Fig. 5.2). Each lung field on an erect CXR is divided into three zones. The upper zone is an area which lies above a horizontal line drawn from the medial end of the second rib anteriorly. The middle zone lies below this and is bordered inferiorly by a line drawn similarly from the fourth rib. The lower zone lies below this. This opaque density is similar in configuration to a tooth which was dislodged during emergency intubation of this patient. Foreign bodies are not as common in adults compared with children. It can occur silently in patients with decreased conscious level. The typical site is in the right main stem bronchus, as this has a more vertical course than the left. An example is seen in this CT (Fig. 5.3). Bronchoscopic removal is the usual initial treatment of choice.

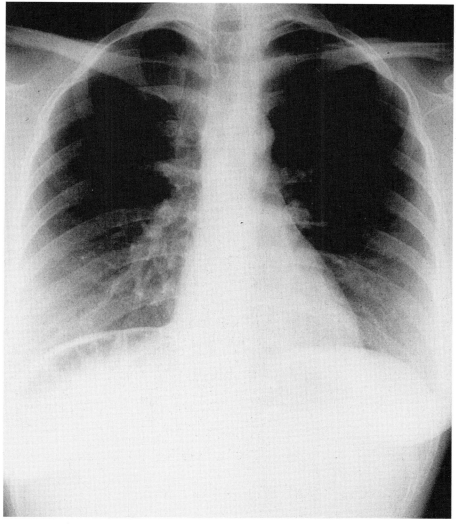

Fig. 6.1

Case 6. This patient was asymptomatic. Her CXR is shown (Fig. 6.1). Name the anomaly.

CASE 6 CHILAIDITI'S SIGN

Chilaiditi described this normal variant in 1911 where the transverse colon is interposed between the right hemidiaphragm and the liver. Its prevalence is thought to be 0.025%. Occasional reports describe patients with Chilaiditi's syndrome where patients complain of intermittent abdominal pain requiring laparotomy to rule out other causes of peritonism, e.g. perforated ulcer, ruptured appendix. The recognition of the haustrations (indicative of large bowel origin) in the bowel shadows is crucial to the diagnosis of Chilaiditi's sign.

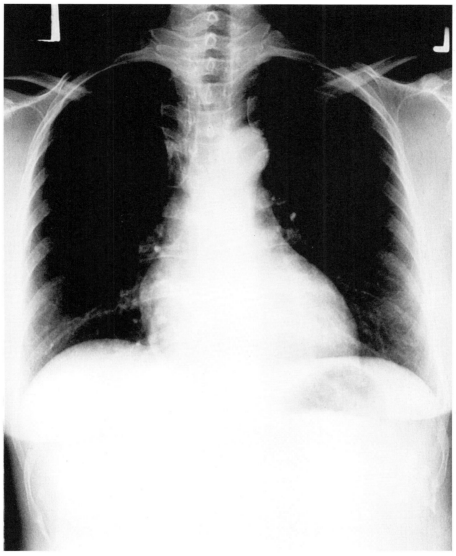

Fig. 7.1

Case 7. This patient was asymptomatic. The CXR is shown (Fig. 7.1).

Fig. 7.2

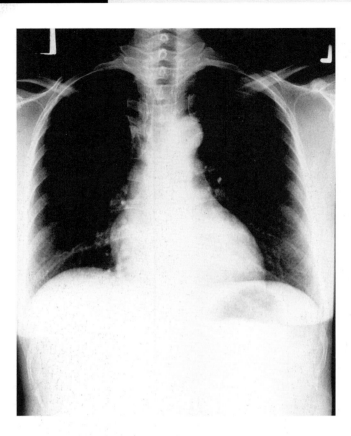

CASE 7 AZYGOUS LOBE

There is a curvilinear density adjacent to the right superior mediastinum with an ovoid lower density at its lower end (the azygous vein). The azygous lobe is the commonest CXR normal variant seen in up to 0.4% of individuals. This is an embryologic variation which results in an accessory lobe at the right upper lobe. The fissure (Fig. 7.2) is due to the invagination of the azygous vein and the condition is of no clinical significance.

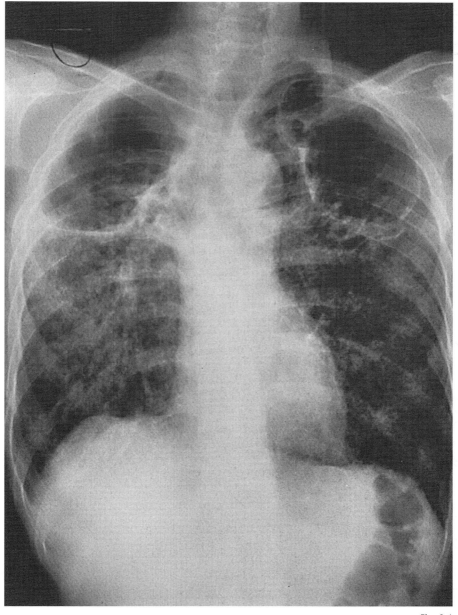

Fig. 8.1

Case 8. This was an 80-year-old male with fever, productive cough, hemoptysis, and loss of weight. This was his CXR (Fig. 8.1). What is the diagnosis?

CASE 8   ACTIVE PULMONARY TUBERCULOSIS

The CXR shows bilateral upper lobe infiltrates with cavities, suggestive of active pulmonary tuberculosis. In general, thin-walled cavities (<5 mm) tend to be infective and, when thick-walled (>10 mm), squamous cell carcinoma of the lung enters into the differential diagnosis. Tuberculosis tends to afflict the upper lobes and apical segment of the lower lobes. However, within the upper lobe, anterior segment involvement is rare. Diagnosis is confirmed by obtaining sputum and staining with fluorochrome or Zeil Nielson and culturing with Lowenstein Jansen media. Cavitary upper lobe disease has good correlation with a sputum positive smear and hence is extremely infectious. Other differential diagnoses of cavitary pulmonary lesions include infections from *Staphylococcus*, *Klebsiella*, anaerobes, and non-infectious causes like squamous cell carcinoma of the lung, pulmonary infarcts, Wegener's granulomatosis, and rheumatoid nodules.

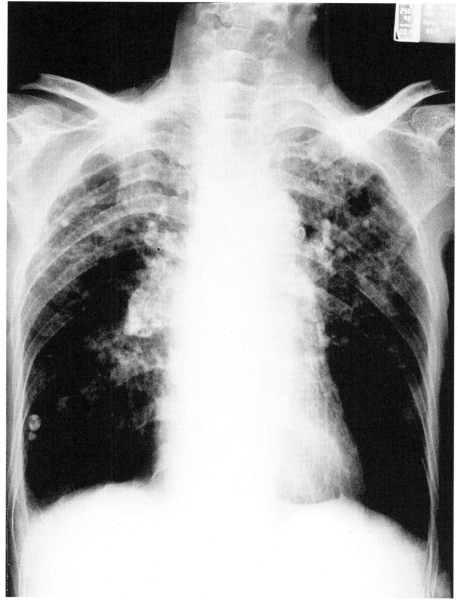

Fig. 9.1

Case 9. This 80-year-old male used to work in a sand quarry. He was asymptomatic. His CXR is shown (Fig. 9.1). What is the diagnosis?

CASE 9 SILICOSIS

The CXR shows bilateral infiltrates and calcified nodules in both upper lobes. Differential diagnoses of upper lobe infiltrates include silicosis, tuberculosis, and ankylosing spondylitis. There is also egg-shell calcification of the hilar lymph nodes. The egg-shell calcification plus the upper lobe nodules are typical of silicosis. Differential diagnoses of egg-shell calcification include sarcoidosis, Hodgkin's lymphoma following radiotherapy, and coal-worker's pneumoconiosis.

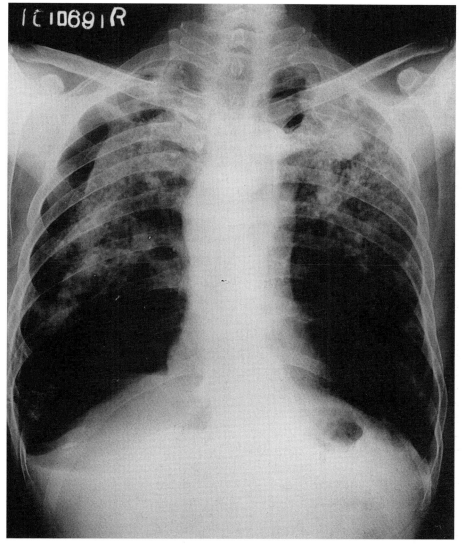

Fig. 10.1

Case 10. This 80-year-old male presented with right-sided chest pain and breathlessness. He gave a long history of exertional dyspnea. The CXR is shown (Fig. 10.1).

CASE 10 SILICOSIS WITH PROGRESSIVE MASSIVE FIBROSIS (PMF)

This patient's CXR shows a right pneumothorax. In addition, there are bilateral diffuse nodules (<10 mm but >2 mm) which could be due to metastatic adenocarcinoma, silicosis, disseminated histoplasmosis, or varicella infection. In silicosis, some nodules may coalesce to form conglomerate masses in the upper lobes called progressive massive fibrosis. Patients with silicosis are predisposed to pulmonary tuberculosis and serial CXR comparison is useful.

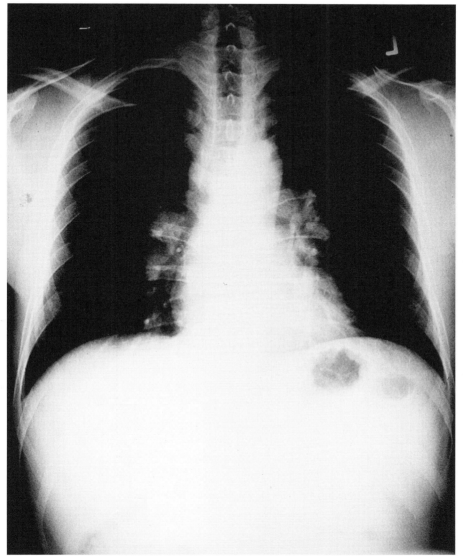

Fig. 11.1

Case 11. This 40-year-old male of African origin was asymptomatic and had a routine CXR (Fig. 11.1). What is the likely diagnosis?

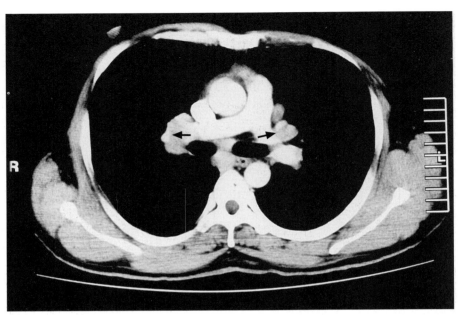

Fig. 11.2

CASE 11 BILATERAL HILAR AND MEDIASTINAL ADENOPATHY FROM SARCOIDOSIS

CXR shows bilateral symmetrically enlarged hilar and mediastinal lymph nodes. CT (Fig. 11.2) confirms this finding, typical of sarcoidosis. The main differential diagnoses would be lymphoma and tuberculosis, but the lymphadenopathy would then be asymmetrical. Bronchoscopy and transbronchial lung biopsy are positive in 60% of cases, showing non-caseating granulomas and culture negative for tuberculosis and fungus. Blind endobronchial biopsies increase the yield by another 20% but the gold standard is mediastinoscopy. Incidence in people of African origin is ten times higher than in Caucasians.

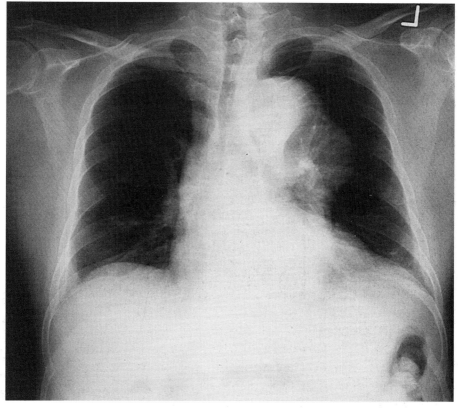

Fig. 12.1

Case 12. A 60-year-old male presented at the Emergency Room with severe chest pain of sudden onset. This was his CXR (Fig. 12.1). What is the diagnosis?

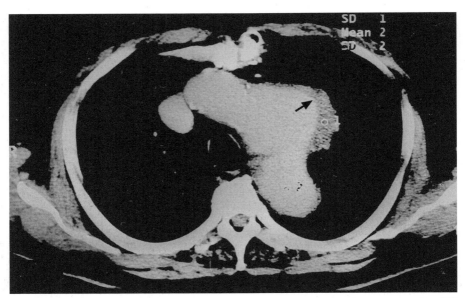

Fig. 12.2

CASE 12 DISSECTING THORACIC ANEURYSM

The CXR shows widening of the superior mediastinum and a well-defined mass inferior and contiguous with the arch of the aorta. In this clinical context, dissection of the arch of the aorta has to be excluded. CT Thorax in another patient shows the presence of an aneurysm (Fig. 12.2) at the aortic arch with thrombus.

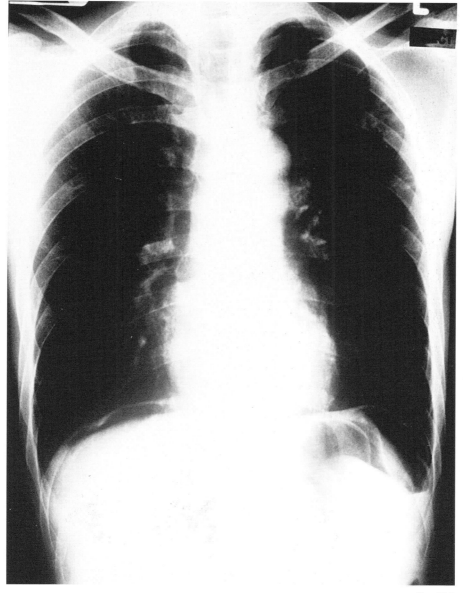

Fig. 13.1

Case 13. This 80-year-old male smoker is a known case of COPD. He pre-
sented with epigastric pain and worsening of shortness of breath.
Arterial blood gas showed acute metabolic acidosis. This was his CXR
(Fig. 13.1). What is the most obvious abnormality?

CASE 13 PNEUMOPERITONEUM DUE TO PERFORATED PEPTIC ULCER

The CXR shows free air under the right hemidiaphragm, in addition to features of hyperinflation. The possibilities include perforated peptic ulcer or GI malignancy, recent laparoscopy/laparotomy, and peritoneal dialysis. It is important to do an erect CXR for the free air to rise to the top of the abdomen. For patients with a nasogastric tube in place, instillation of 200 ml of free air before the CXR may aid the diagnosis.

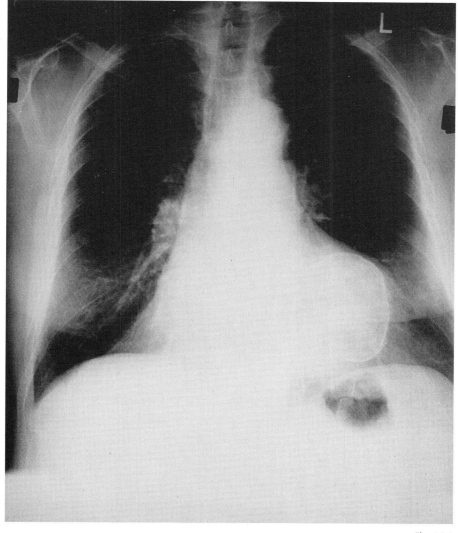

Fig. 14.1

Case 14. This 75-year-old male had a history of myocardial infarction and now presented with recurrent Ventricular Tachycardia. These were his CXR, PA and lateral (Figs. 14.1 and 14.2).

Fig. 14.2

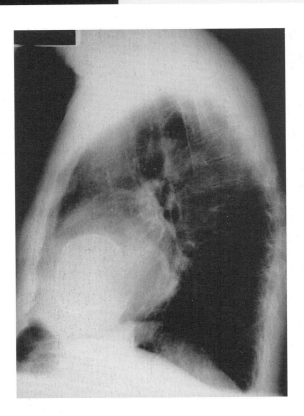

CASE 14 CALCIFIED LEFT VENTRICULAR ANEURYSM

The PA and lateral CXR confirm an arcuate density in the region of the left ventricle. This is typical of calcification of a left ventricular aneurysm, usually secondary to previous myocardial infarction. Surgical resection of the aneurysm is potentially curative.

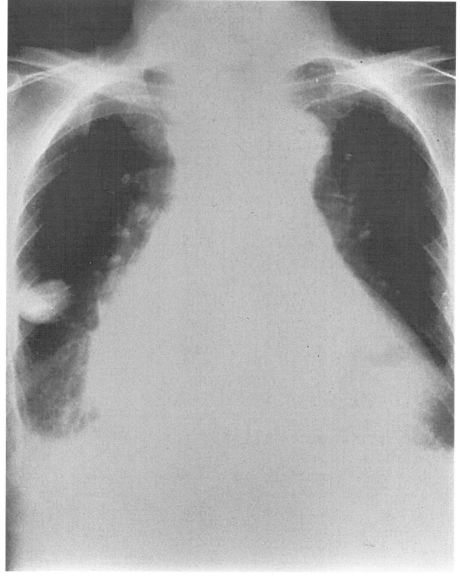

Fig. 15.1

Case 15. A 60-year-old male presented with exertional dyspnea, orthopnea, paroxysmal nocturnal dyspnea, and bilateral painless ankle swelling. This was his CXR (Fig. 15.1). What is the abnormality and subsequent management?

Fig. 15.2

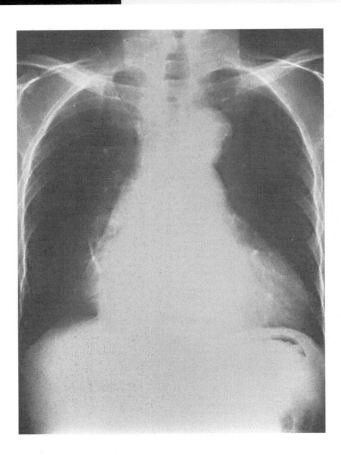

CASE 15 PSEUDOTUMOR DUE TO LOCULATED RIGHT PLEURAL EFFUSION

The CXR shows classic evidence of congestive heart failure with cardiomegaly, upper lobe venous diversion, and bilateral pleural effusions. In addition, there is an ovoid mass in the right middle zone which seems to be related to the transverse fissure. This is typical of a pseudotumor due to a loculated pleural effusion distending the transverse fissure. Appropriate management would include diuretics and treatment of the cardiac failure. Repeat CXR a week later showed the disappearance of the pseudotumor (Fig. 15.2).

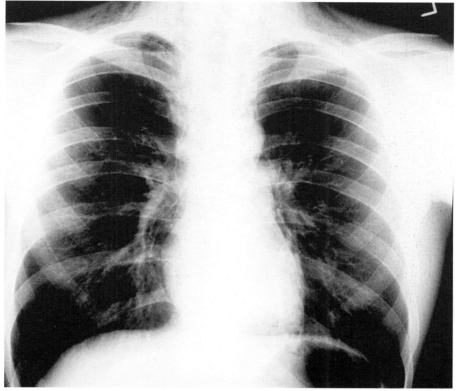

Fig. 16.1

Case 16. A 30-year-old male was seen in the Emergency Room for acute onset chest pain. This was his CXR (Fig. 16.1). Name the most obvious abnormality.

Fig. 16.2

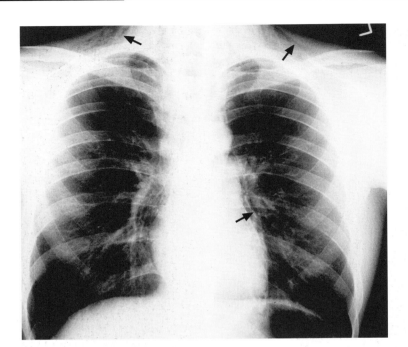

CASE 16 MEDIASTINAL EMPHYSEMA (PNEUMOMEDIASTINUM)

The CXR shows free air in the mediastinum and subcutaneous tissues of the neck
(Fig. 16.2). The mediastinal air could have come from disruption of the integrity of
the lung, major airways, or the esophagus. A history of trauma (e.g. motor vehicle
accident with blunt injury to the anterior chest wall by the steering wheel) or iatro-
genic instrumentation (e.g. recent endoscopy) is important. Descending infections
by gas-producing organisms from the oral cavity and neck can cause severe mediastinitis and result in a similar appearance.

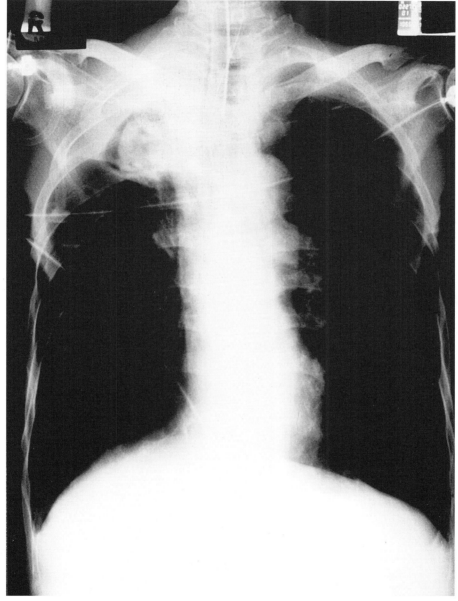

Fig. 17.1

Case 17. An 80-year-old male presented with massive hemoptysis and was intubated. This was his CXR (Fig. 17.1). He gave a past history of being treated for tuberculosis many years ago.

CASE 17 MYCETOMA RIGHT UPPER LOBE

The CXR shows a right upper lobe ball within a cavity (air crescent sign) pathogmonic of a mycetoma (also called aspergilloma). A lateral decubitus X-ray may demonstrate the fungal ball shifting position. In this condition, a preformed cavity becomes colonized, usually by the fungus *Aspergillus fumigatus*. Cavitary disease may be secondary to fibrotic lung disease, e.g. previous tuberculosis, sarcoidosis, or ankylosing spondylitis. Massive hemoptysis can result and bronchial angiogram with embolotherapy (using coils or gel foam) is temporizing. Surgical resection is definitive, but bronchopleural fistula may result. Unfortunately, most patients have insufficient pulmonary reserve to allow surgical resection.

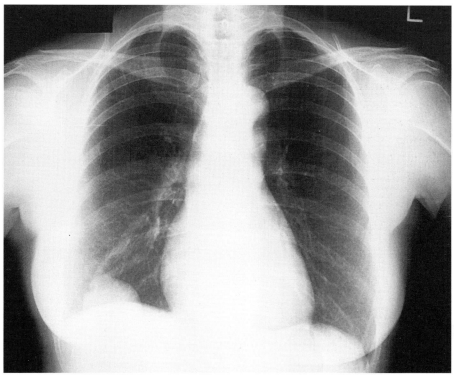

Fig. 18.1

Case 18. This 68-year-old female had recurrent epistaxis. This was her CXR (Fig. 18.1). What is the diagnosis?

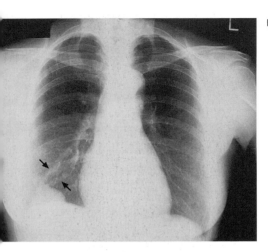

Fig. 18.2

Fig. 18.3

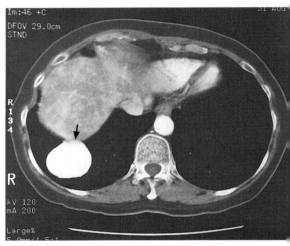

CASE 18 HEREDITARY HEMORRHAGIC TELANGIECTASIA OR OSLER WEBER RENDU DISEASE

The CXR shows a mass in the right lower zone. The mass has a sharp margin and two vessels (supplying artery and draining vein) leading to the mass (Fig. 18.2). The CT (Fig. 18.3) shows marked enhancement of the "mass" with contrast confirming the presence of pulmonary arteriovenous malformation (pAVM). Of patients with pAVM, 60% have Osler's disease, and 10% of patients with Osler's disease have pAVM. This condition is autosomal dominant. Other sites of involvement include skin, nose (epistaxis), gastrointestinal (GI) system (bleeding GI and anemia). Paradoxical embolism can occur resulting in cerebral vascular accidents or brain abscess. Pulmonary angiogram and embolotherapy are recommended if the pAVM is more than 2 mm.

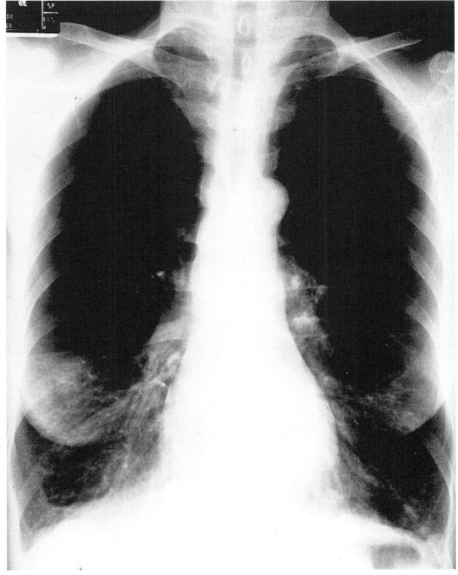

Fig. 19.1

Case 19. An 80-year-old female, 100-pack-a-year smoker with 5-year history of dyspnea on exertion. Describe her CXR (Fig. 19.1). What is the diagnosis?

CASE 19 CHRONIC OBSTRUCTIVE PULMONARY DISEASE (COPD)

The CXR of COPD typically demonstrates evidence of air trapping. The signs are horizontality of the ribs, hyperinflated lungs (normally the right sixth rib bisects the right hemidiaphragm), hyperlucent lung fields, bilateral symmetrical attenuated pulmonary vasculature, long tubular heart, scalloping and flattening of the diaphragm. The commonest cause of COPD worldwide is tobacco smoking. However, it is recognized that alpha-1-antitrypsin deficiency can also cause emphysema. One should look out for alpha-1-antitrypsin deficiency, especially if the COPD patient is young (<45 years old) or demonstrates basal predominance on CXR.

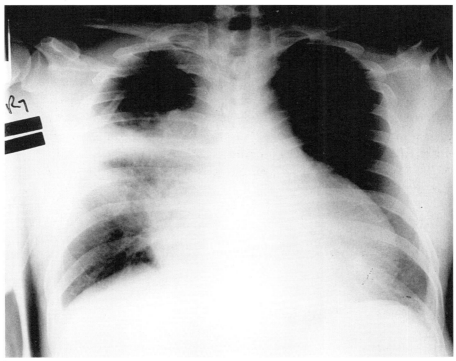

Fig 20.1

Case 20. This 55-year-old male was admitted in shock. He was recently diagnosed with inoperable lung cancer. Clinical exam also showed distended neck veins and muffled heart sounds. This was his CXR (Fig. 20.1). What is the diagnosis?

CASE 20 CARDIAC TAMPONADE FROM MASSIVE PERICARDIAL EFFUSION

Beck described a triad of hypotension, muffled heart sounds, and elevated jugular venous pressure due to cardiac tamponade from pericardial effusion. Immediate pericardiocentesis is life-saving. The common causes of pericardial effusion include malignancy, congestive heart failure, tuberculosis, systemic lupus erythematosus, Dressler's syndrome, and uremia. This CXR shows a globular enlargement of the heart, typical of a large pericardial effusion. In addition, there is a mass in the right lung in keeping with the primary lung cancer.

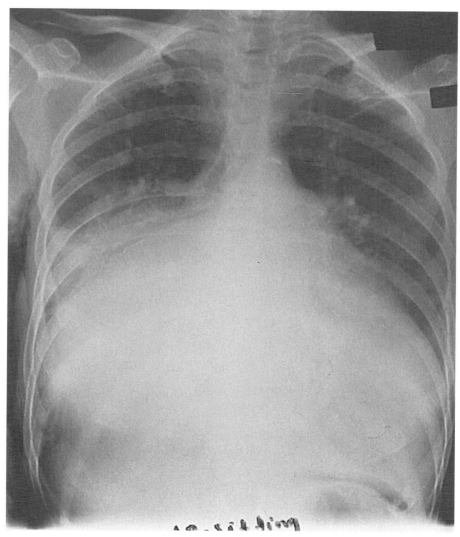

Fig 21.1

Case 21. This 65-year-old male had a long history of dyspnea on exertion, orthopnea, and bilateral ankle edema. This was his CXR (Fig. 21.1). Should a thoracocentesis be done?

Fig. 21.2

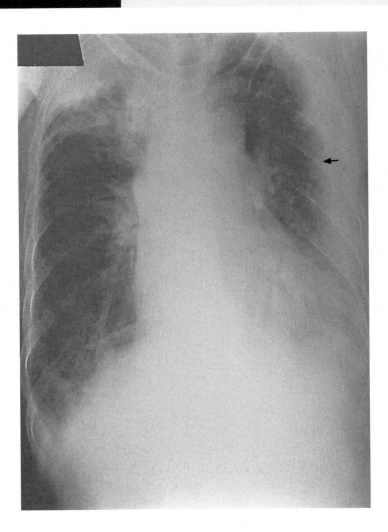

CASE 21 SEVERE CARDIOMEGALY DUE TO END STAGE VALVULAR HEART DISEASE

The CXR shows very severe cardiomegaly (the normal cardiothoracic ratio is defined as less than 0.5). Both costophrenic angles show lucency due to aerated lung, making it unlikely that the patient has massive pleural effusions. The carina is also splayed indicating an enlarged left atrium due to severe mitral valve disease. Hence, in this patient, thoracocentesis should not be done. A simple way to confirm the presence of a pleural effusion is to take a lateral decubitus CXR. A free-flowing effusion will layer out (Fig. 21.2). However, the absence of layering on a lateral decubitus CXR does not preclude the presence of a significant pleural effusion as it may be loculated due to an empyema.

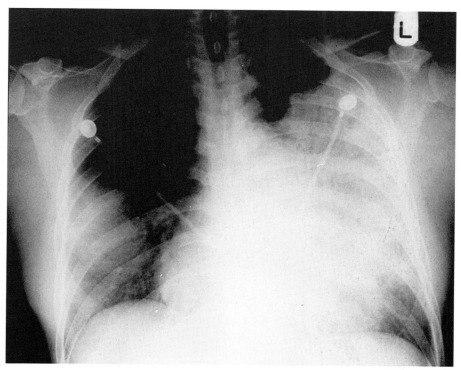

Fig. 22.1

Case 22. This 75-year-old female presented with acute respiratory failure. She had been sick for two weeks with fever, cough, and purulent sputum. This was her CXR (Fig. 22.1). What is the diagnosis?

CASE 22 SEVERE PNEUMONIA

See Case 1. The CXR shows opacities with air bronchograms involving both lung fields. This is typical of severe pneumonia as evidenced by multilobar involvement. Typical organisms include *Streptococcus pneumoniae*, *Legionella*, and gram negatives like *Klebsiella* and *Pseudomonas aeroginosa*. In South-East Asia, another possible etiologic agent is *Burholderia pseudomallei* (Meliodosis). Treatment will require combination parenteral antibiotics, usually beta lactams plus macrolide or fluoroquinolone. The prognosis is dependent not just upon the severity of presentation but also underlying age and co-morbidities, e.g. cancer, heart, liver, or renal disease, and stroke. This patient's pneumonia was confirmed to be due to severe Legionellosis.

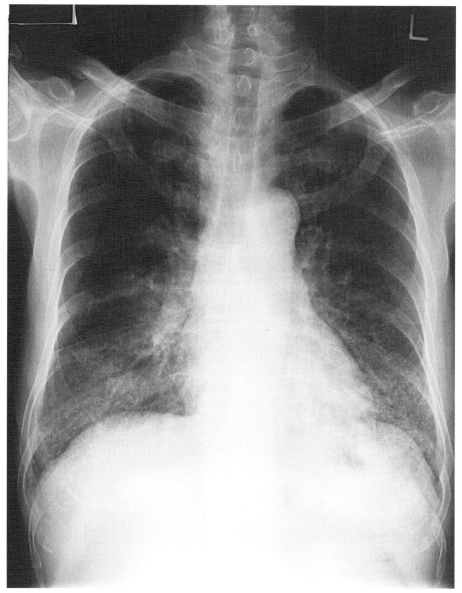

Fig. 23.1

Case 23. A 30-year-old male presented with cough, shortness of breath and loss of weight over four months. This was his CXR (Fig. 23.1). What is the most likely diagnosis? What physical sign would be useful?

CASE 23 *PNEUMOCYSTIS CARINII* PNEUMONIA (PCP)

The CXR shows bilateral infiltrates and air bronchograms with a perihilar distribution. The heart size is normal. There are no Kerley B lines or evidence of upper lobe venous diversion. All these are typical features of PCP. PCP is the most common life-threatening opportunistic infection in HIV disease. Generally, the most common opportunistic infection in HIV is oral candidiasis. Oral candidiasis should be looked for in any young patient with pneumonia as it may be a sign of T-cell immune deficiency. PCP can be diagnosed by sputum induction or bronchoalveolar lavage. Note that 10% of PCP patients could have a normal CXR.

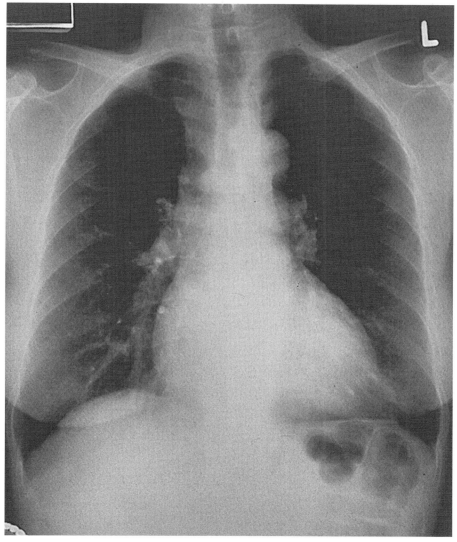

Fig. 24.1

Case 24. This middle-aged female non-smoker was recently diagnosed and treated as for asthma with little response. This was her CXR (Fig. 24.1). What is the diagnosis?

Fig. 24.2

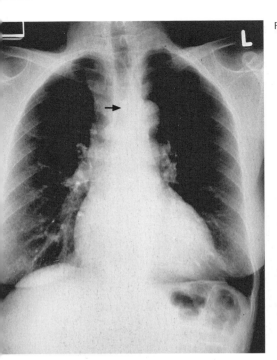

Fig. 24.3

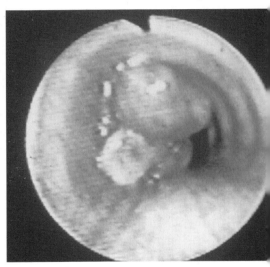

CASE 24 TRACHEAL TUMOR DUE TO ADENOID CYSTIC CARCINOMA

All patients diagnosed with asthma should have a CXR. In addition to looking for pneumothorax and transient pulmonary infiltrates, one should pay attention to the tracheal air column. Any obstruction to the major airway can produce a wheeze. If the obstruction is high up, i.e. extrathoracic, the sound is described as stridor, i.e. during inspiration. This is in contradistinction to rhonchi which is classically expiratory and due to small airway obstruction. The CXR here shows a bulge in the lateral wall of the mid-trachea (Fig. 24.2) due to a tumor. Possibilities include squamous cell carcinoma, metastases, mucoepidermoid carcinoma, adenoid cystic carcinoma and carcinoid tumor. Flexible bronchoscopy in this patient showed a mid-tracheal tumor (Fig. 24.3) and biopsy showed adenoid cystic carcinoma (a low-grade malignancy).

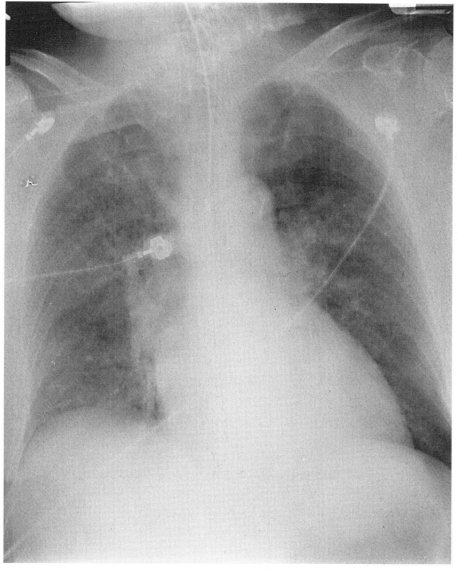

Fig. 25.1

Case 25. This was a routine CXR (Fig. 25.1) in an ICU patient who was admitted for aspiration pneumonia. Name the most obvious abnormality.

Fig. 25.2

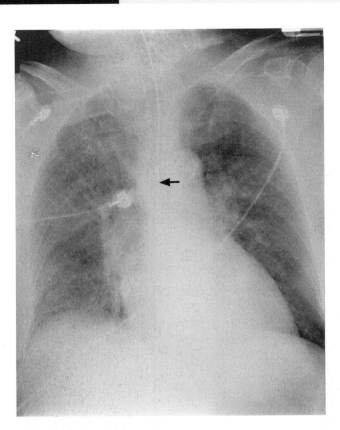

CASE 25 MALPOSITIONED NASOGASTRIC TUBE

The tip of the nasogastric tube should be seen within the gastric bubble. In this case, the tube has coiled at the esophageal cardia and ended up in the mid-esophagus (Fig. 25.2). Feeding within the esophagus may result in fatal aspiration. The CXR also shows evidence of right lower lobe infiltrates, a typical site for aspiration pneumonia.

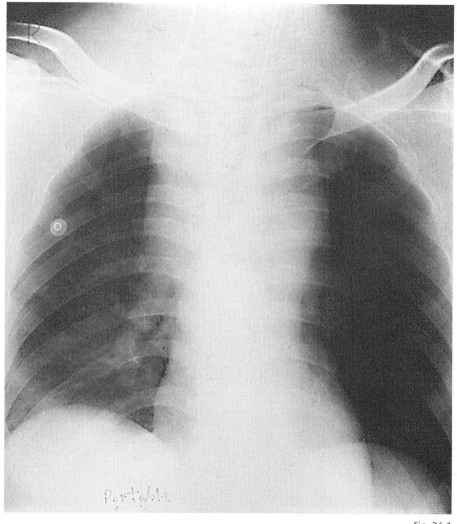

Fig. 26.1

Case 26. This was a routine CXR (Fig. 26.1) taken after placement of a subclavian central venous catheter.

CASE 26 MALPOSITIONED RIGHT CENTRAL VENOUS CATHETER

The most obvious abnormality is that the right subclavian central venous catheter tip has curled upwards into the right internal jugular vein instead of downwards into the superior vena cava. The other finding is that of soft tissue swelling in the right neck and superior mediastinal widening. This patient had severe coagulopathy and repeated attempts at the central venous catheter insertion resulted in a neck hematoma which had also tracked inferiorly causing a mediastinal hematoma. As a result, the patient required intubation to secure the airway.

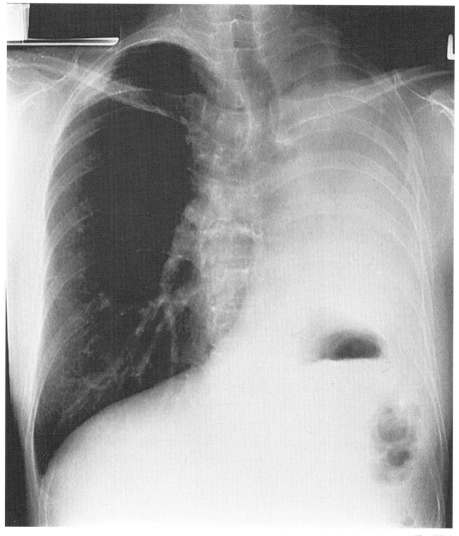

Fig. 27.1

Case 27. This patient was asymptomatic. Past history was significant for previous thoracotomy. The CXR is shown (Fig. 27.1).

Fig. 27.2

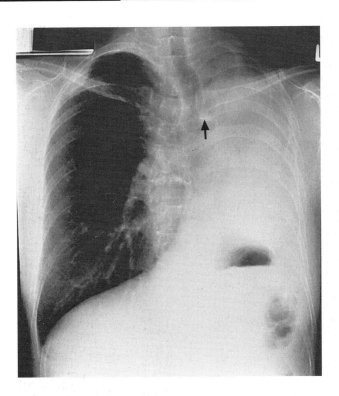

CASE 27 POST LEFT PNEUMONECTOMY

There is a homogenous whiteout of the left hemithorax. The differential diagnoses are complete left lung collapse or post left pneumonectomy. The elevation of the gastric bubble and leftward shift of mediastinum here rule out a massive left pleural effusion. The presence of surgical clips in the left hemithorax in the vicinity of the left main-stem bronchus (Fig. 27.2) makes a left pneumonectomy very likely.

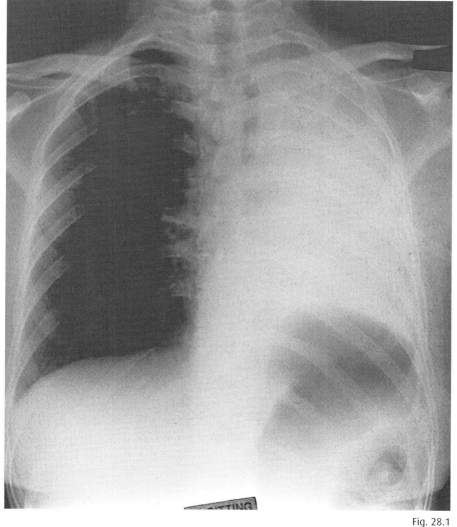

Fig. 28.1

Case 28. This patient presented with recent onset of dyspnea and streaky hemoptysis. The CXR is shown (Fig. 28.1). What is the radiological diagnosis?

Fig. 28.2

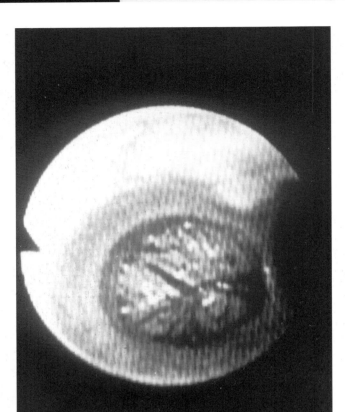

CASE 28 COLLAPSE/ATELECTASIS OF THE LEFT LUNG

See Case 27. There is a homogenous whiteout of the left hemithorax. As in the previous case, there is evidence of volume loss in the left lung with shift of mediastinum to the left, crowding of the left-sided ribs and elevation of the left hemidiaphragm. Flexible bronchoscopy demonstrated near-total occlusion of the left main-stem bronchus by a tumor (mucoepidermoid carcinoma, Fig. 28.2). Laser resection of the tumor was then performed, resulting in restoration of ventilation to the left lung.

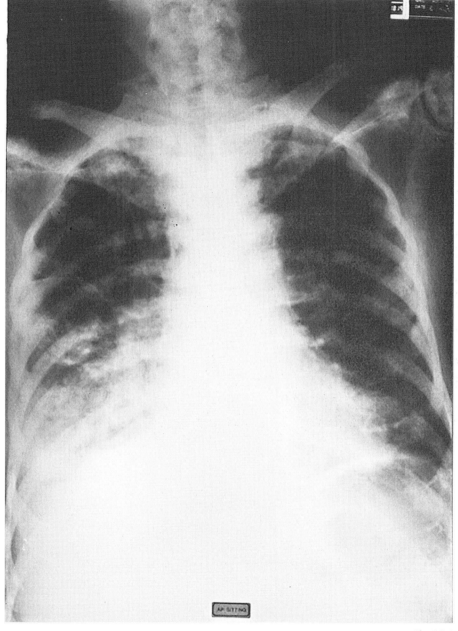

Fig. 29.1

Case 29. This elderly male patient had recent loss of weight and bone pains. What is the most obvious CXR abnormality (Fig. 29.1)? Name the differential diagnoses?

CASE 29 INCREASED BONY DENSITIES DUE TO OSTEOSCLEROTIC METASTASES

The bones show patchy increased density due to metastases from carcinoma of the prostate. The differential diagnoses are Paget's disease and Fluorosis. Cancer of breast or lymphoma may also cause the same appearance. The CXR also shows right lower lobe infiltrates, suggesting aspiration pneumonia, common in the last stages of patients debilitated with cancer.

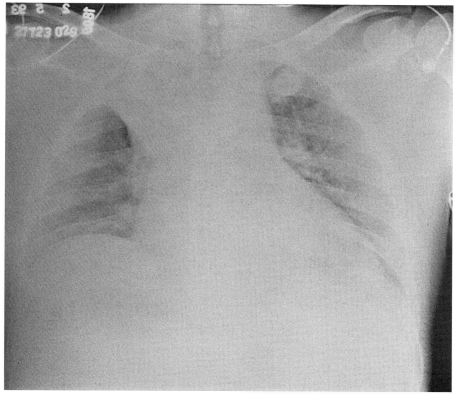

Fig. 30.1

Case 30. This elderly male had recent onset of streaky hemoptysis.
Name the radiological sign (Fig. 30.1).

Fig. 30.2

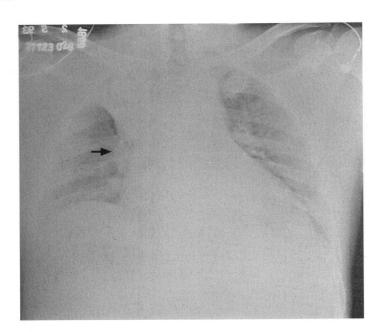

CASE 30 GOLDEN'S S SIGN OF RIGHT UPPER LOBE COLLAPSE

There is a homogeneous density in the right upper zone and elevation of the transverse fissure. Instead of the transverse fissure being straight, there is a bulge at the medial end (Fig. 30.2), giving it an inverted S shape. Golden described this sign and the explanation for it is that the upper lobe collapse is due to a right hilar mass which accounts for the medial bulge.

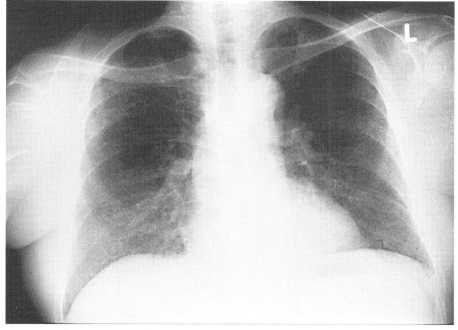

Fig. 31.1

Case 31. This diabetic presented with prolonged pyrexia of uncertain origin (PUO). Describe the CXR abnormality (Fig. 31.1).

CASE 31 DIFFUSE MILIARY SHADOWS DUE TO MILIARY TUBERCULOSIS

CXR shows bilateral diffuse miliary shadows (<2 mm diameter) due to miliary tuberculosis. The differential diagnoses include previous varicella infection, disseminated histoplasmosis, and silicosis. A travel history to endemic countries or a relevant occupational history is helpful to distinguish the various causes. Another very rare cause of such a CXR pattern is pulmonary alveolar microlithiasis.

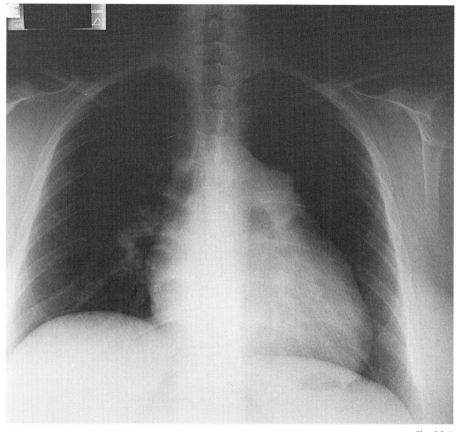

Fig. 32.1

Case 32. This 25-year-old female had tiredness and shortness of breath for the past year. Describe the CXR (Fig. 32.1).

CASE 32 PRIMARY PULMONARY HYPERTENSION

This patient fits the typical clinical and radiological profile of a patient with primary pulmonary hypertension. The pulmonary arteries are markedly enlarged with the right atrial chamber also enlarged. The normal right pulmonary descending artery diameter is less than 16 mm in males and 15 mm in females. The lung fields are clear and the lung volumes normal making lung disease causing pulmonary hypertension unlikely. Other causes to be ruled out are congenital heart disease and chronic pulmonary thromboembolism.

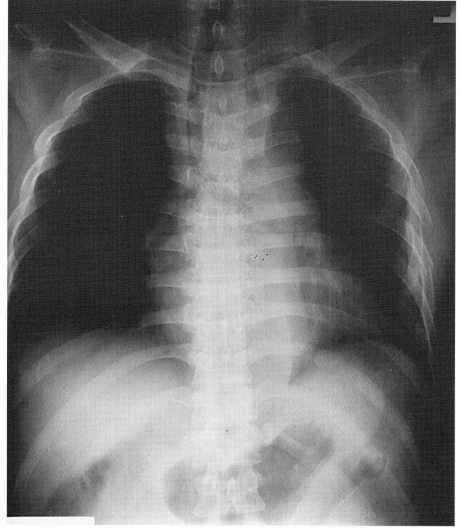

Fig. 33.1

Case 33. This middle-aged male was involved in a motor vehicle accident where he was the driver and his vehicle was hit from behind resulting in intense chest pain. His CXR is shown (Fig. 33.1).

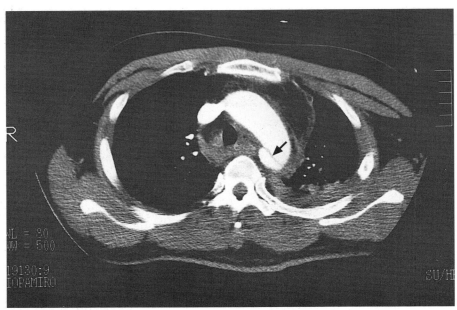

Fig. 33.2

CASE 33 TRAUMATIC AORTIC DISRUPTION

This CXR shows evidence of a widened superior mediastinum and loss of the aortic knuckle and obliteration of the aorto-pulmonary window. There is left apical capping as a result of mediastinal blood tracking to the extrapleural region of the left hemithorax. The trachea is deviated to the right and the left main-stem bronchus is depressed. The fifth and sixth ribs on the left side are fractured. Sometimes there is an associated left hemothorax. All these are typical features of traumatic aortic disruption, which usually occurs just distal to the ligamentum arteriosum (Fig. 33.2).

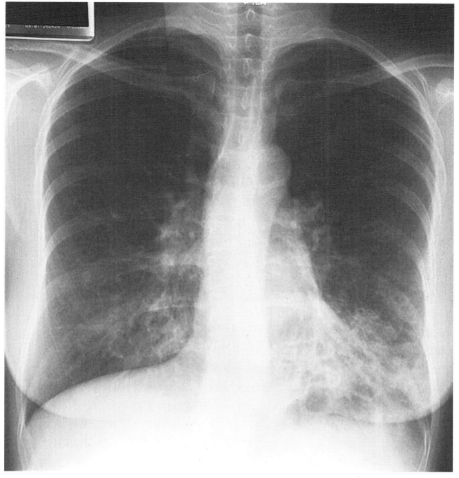

Fig. 34.1

Case 34. This middle-aged female had chronic productive cough for many years. What is the diagnosis (Fig. 34.1)?

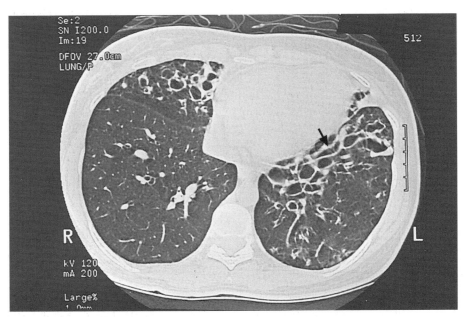

Fig. 34.2

CASE 34 BRONCHIECTASIS AFFECTING BOTH LOWER LOBES

The CXR shows infiltrates especially in the right middle lobe and the left lower lobe. The ring shadows and tramlines indicate the presence of dilated and thickened airways. The CXR findings were noted a few years previously indicating its chronicity. The accepted modality for the diagnosis of bronchiectasis is a high-resolution CT Thorax which demonstrates these dilated airways in the left lower lobe (Fig. 34.2) using very thin (1–2 mm) slices. Bronchography is now seldom used.

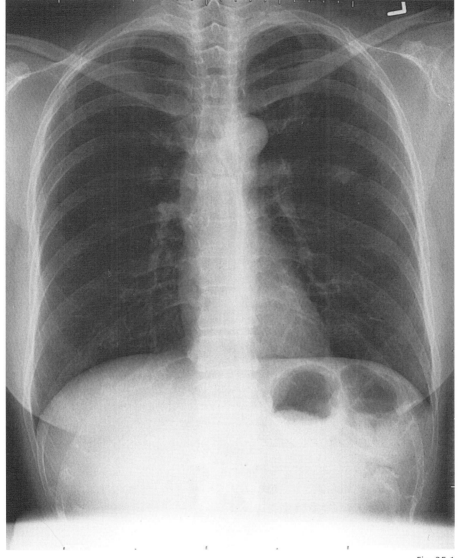

Fig. 35.1

Case 35. This middle-aged female smoker was asymptomatic. Describe the CXR abnormality (Fig. 35.1).

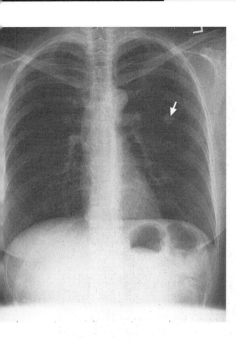

Fig. 35.2

Fig. 35.3

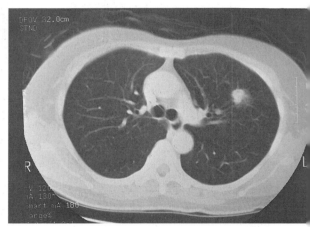

CASE 35 SOLITARY PULMONARY NODULE (SPN) DUE TO PRIMARY LUNG CANCER

The CXR shows a 1.5 cm solitary pulmonary nodule in the left upper lobe (Fig. 35.2). An SPN is described as a single nodule (less than 4 cm) surrounded by normal lung parenchyma. The differential diagnoses for SPN include pseudo nodules (e.g. skin tags, nipple shadows, and bone lesions), primary lung cancer, solitary metastases, granulomas, arteriovenous malformations, pseudo tumors, and hamartomas. In this patient, the CXR a year ago did not demonstrate the shadow. CT (Fig. 35.3) also demonstrates the nodule to be non-calcified and the margins show spiculation making the nodule highly suspicious for malignancy. Thoracotomy and lung biopsy showed primary Stage 1 lung cancer (adenocarcinoma).

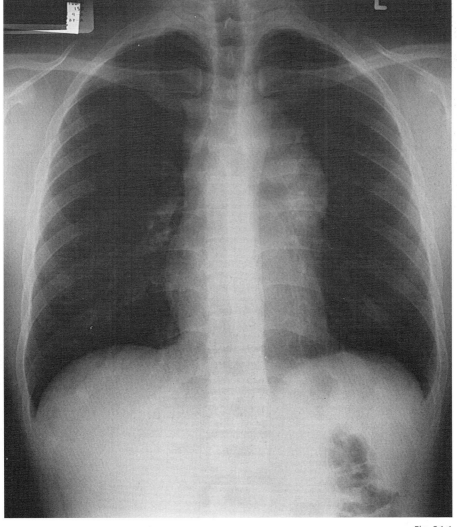

Fig. 36.1

Case 36. This middle-aged male had loss of weight and bilateral cervical lymphadenopathy. His CXR is shown (Fig. 36.1).

Fig. 36.2

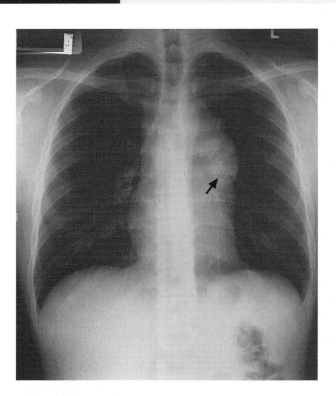

CASE 36 MEDIASTINAL LYMPHADENOPATHY DUE TO LYMPHOMA

See Case 11. The CXR shows asymmetric distortion of the mediastinal contour by markedly enlarged lymph nodes overlying the left hilum. This is described as the hilar overlay sign – the normal left pulmonary artery (Fig. 36.2) is seen through the mass (lying at the anterior mediastinum). Other differential diagnoses include chronic lymphocytic leukemia, sarcoidosis, Castleman's disease, and granulomatous disease like tuberculosis or histoplasmosis. The histology from mediastinoscopy in this patient showed Non Hodgkin's lymphoma.

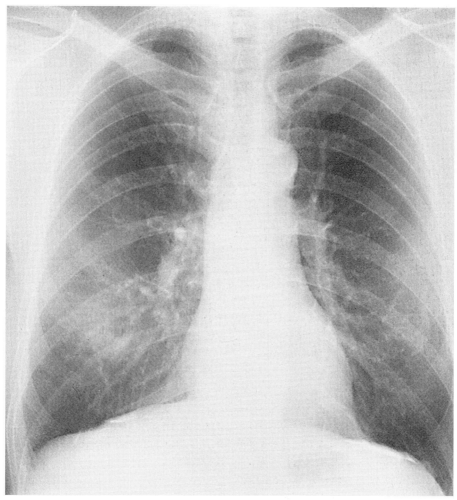

Fig. 37.1

Case 37. This elderly male was asymptomatic. What is the abnormality on his CXR (Fig. 37.1)? What is the cause?

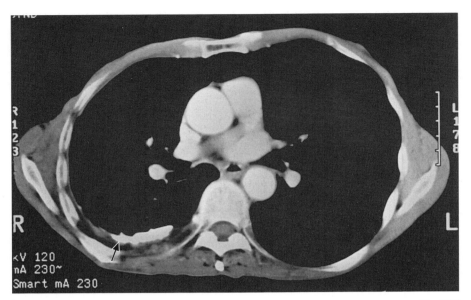

Fig. 37.2

CASE 37   BILATERAL CALCIFIED PLEURAL PLAQUES
DUE TO ASBESTOS EXPOSURE

The CXR shows bilateral calcified pleural plaques, especially over the diaphragmatic pleura. The mid-zones show *en face* calcification (holly leaf sign). This is typical of asbestos exposure. Previously asbestos was commonly used as an insulating material. Asbestos exposure can also result in benign pleural effusion, round atelectasis, pulmonary fibrosis (asbestosis), or malignant mesothelioma. Differential diagnosis of pleural calcification includes previous hemothorax, empyema, and tuberculosis. CT also demonstrates the calcified pleural plaques (Fig. 37.2).

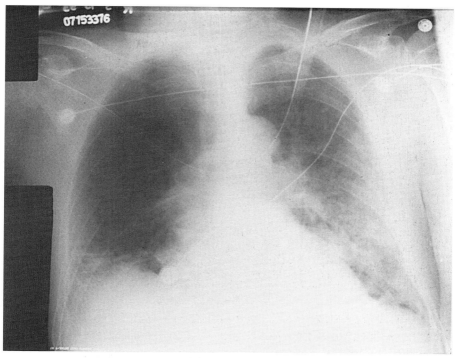

Fig. 38.1

Case 38. This elderly male was bed-bound because of a massive stroke. Over the past week, he developed a low-grade fever and became tachypneic and hypotensive, requiring resuscitation and mechanical ventilation. What is the radiological sign (Fig. 38.1)? What is the diagnosis?

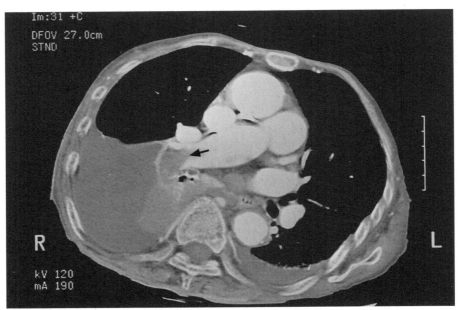

Fig. 38.2

CASE 38 WESTERMARK'S SIGN OF ACUTE PULMONARY EMBOLISM

The CXR shows an oligemic right upper lobe (Westermark's sign) due to acute pulmonary embolism. Other causes of a hyperlucent lung include a right pneumothorax or huge bullae. Other radiological signs of pulmonary embolism are wedge-shaped infarct (Hampton's hump), plate atelectasis, enlarged pulmonary arteries, or small pleural effusion. The CXR may also be normal. CT confirms the clot in the right main pulmonary artery (Fig. 38.2).

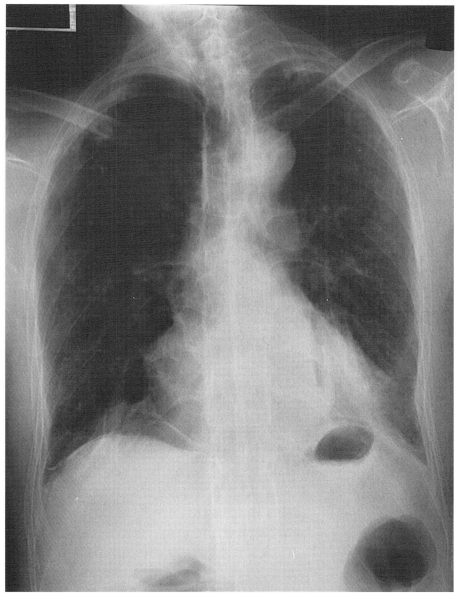

Fig. 39.1

Case 39. This middle-aged male was asymptomatic. What is the CXR abnormality (Fig. 39.1)?

Fig. 39.2

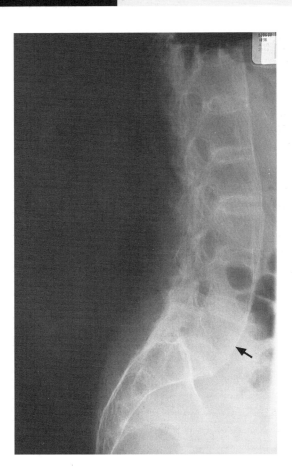

CASE 39 BAMBOO SPINE APPEARANCE DUE TO ANKYLOSING SPONDYLITIS

The most obvious finding is calcification of the interspinous ligaments causing a bamboo spine appearance on CXR, typical of ankylosing spondylitis. This disorder typically affects young males with predominant involvement of the axial spine and the sacroiliac joints (Fig. 39.2). Upper lobe fibrosis may also result. The lung function abnormality that results is usually restrictive. There is a very strong association with HLA-B27.

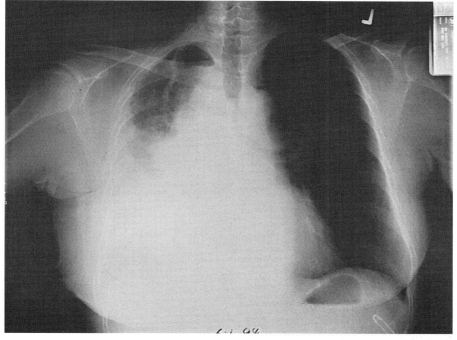

Fig. 40.1

Case 40. This middle-aged female smoker had hemoptysis and loss of weight. What is the CXR abnormality (Fig. 40.1)?

CASE 40 MASS IN RIGHT LUNG ARISING FROM TRACHEA AND RIGHT BRONCHIAL TREE

The CXR shows a mass in the right upper zone with a pleural effusion, suggestive of advanced lung cancer. The lower end of the tracheal air column also shows narrowing, indicating involvement by the cancer. Lung cancer is the commonest cause of malignant pleural effusion and is usually secondary to smoking.

Squamous cell and small cell lung cancer tend to involve the central airways, the latter often associated with mediastinal lymphadenopathy. Adenocarcinoma of the lung tends to present as peripheral nodules.

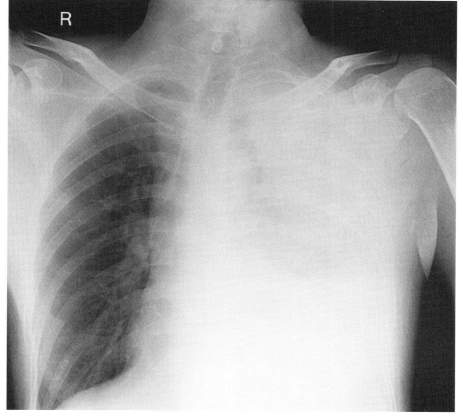

Fig. 41.1

Case 41. This middle-aged male presented with fever, productive cough, and shortness of breath of two weeks' duration. This was his CXR (Fig. 41.1).

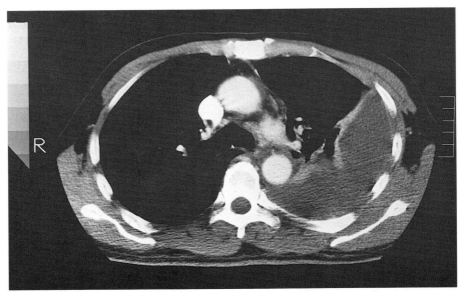

Fig. 41.2

CASE 41 MASSIVE LEFT PLEURAL EFFUSION

The CXR shows a dense homogeneous whiteout of almost the entire left hemithorax associated with a shift of mediastinum to the right, consistent with a massive left pleural effusion. Collapse and previous pneumonectomy may cause a similar appearance except that the mediastinum is shifted to the ipsilateral side. All patients with unilateral pleural effusion should be considered for thoracocentesis to determine the cause of the effusion. The commonest cause of a massive pleural effusion is involvement from lung cancer. In this patient, thoracocentesis yielded frank pus due to an empyema. CT thorax (Fig. 41.2) shows enhancement of both the parietal and visceral pleura, also called the split pleura sign. This results from intense inflammation of the pleura.

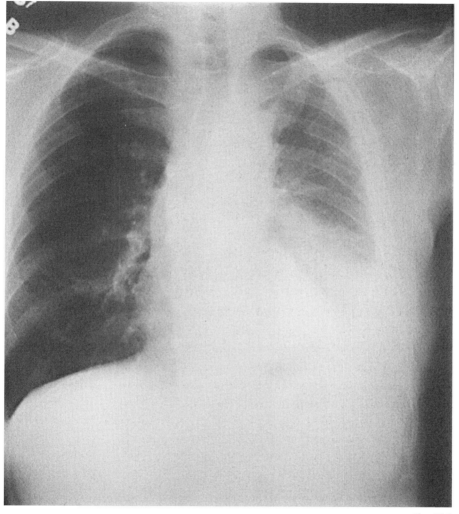

Fig. 42.1

Case 42. This elderly male presented with left-sided persistent chest pain and loss of weight for the past few months. He used to work as an electrician on-board a ship for many years. This was his CXR (Fig. 42.1).

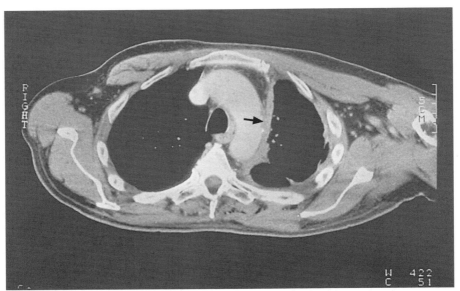

Fig. 42.2

CASE 42 MALIGNANT MESOTHELIOMA

The CXR shows a small left pleural effusion with blunting of the left costophrenic angle. The left hemithorax is smaller than the right. The mediastinum is also widened due to tumor creeping along the pleura. All these are features of malignant mesothelioma, which is a primary malignancy of the pleura and typically spreads along the pleura as demonstrated on CT (Fig. 42.2).

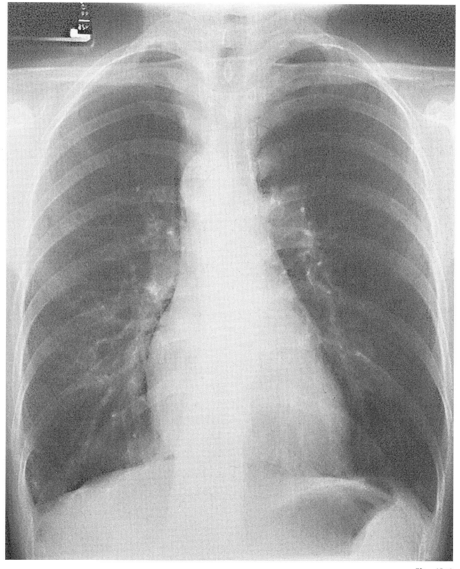

Fig. 43.1

Case 43. This patient was asymptomatic. This was his CXR (Fig. 43.1).

Fig. 43.2

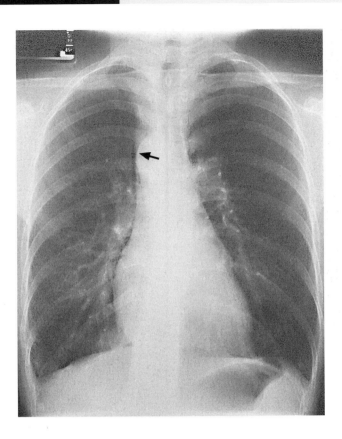

CASE 43 RIGHT-SIDED AORTIC ARCH

The aortic knuckle, which is usually on the left, is now on the right (Fig. 43.2). This is a congenital abnormality. The commonest type is associated with an aberrant anterior left common carotid artery and a retroesophageal left subclavian artery. This is seen in about 1 in 2500 patients and is not associated with any congenital heart disease.

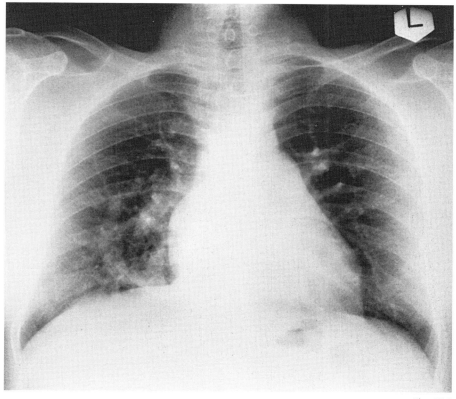

Fig. 44.1

Case 44. This patient was asymptomatic. The CXR is shown (Fig. 44.1).

Fig. 44.2

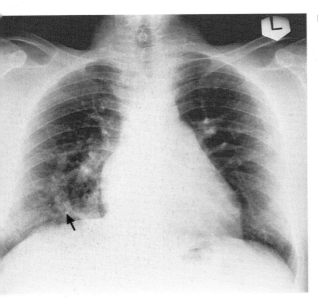

Fig. 44.3

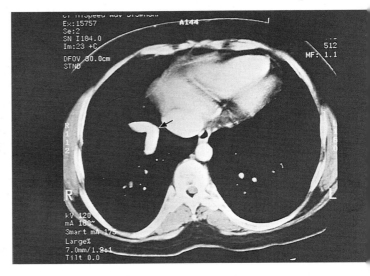

CASE 44 ANOMALOUS PULMONARY VENOUS DRAINAGE – SCIMITAR SIGN

The curvilinear shadow in the right lower zone is called a Scimitar sign. This is due to aberrant drainage of the right inferior pulmonary vein (Fig. 44.2) into the inferior vena cava. This is a congenital anomaly and is usually associated with a small ipsilateral hemithorax and a small or hypoplastic pulmonary artery. This condition is usually of no clinical significance. The CT scan shows the enhancing vein (Fig. 44.3).

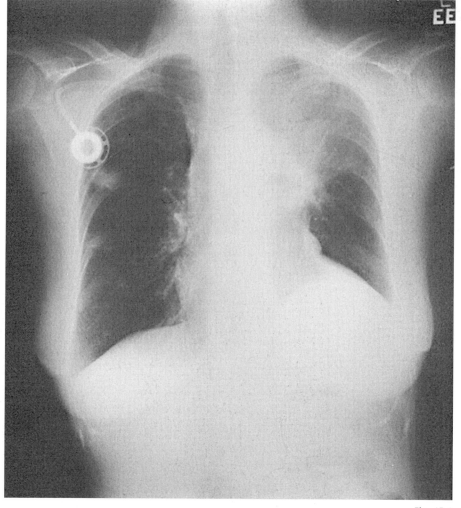

Fig. 45.1

Case 45. This middle-aged female complained of hemoptysis and loss of weight of two months' duration. This was her CXR (Fig. 45.1).

CASE 45 LEFT UPPER LOBE COLLAPSE DUE TO LUNG CANCER

The CXR shows evidence of left upper lobe collapse. There is a hazy, veil-like opacification in the left upper lobe, which does not have a sharp inferior margin unlike right upper lobe collapse (see Case 30). This is because there is usually no left transverse fissure and the lobe collapses anteriorly. There is also volume loss in the left hemithorax as evidenced by an elevated left hemidiaphragm and crowding of the left upper ribs. Sometimes the trachea may also be deviated to the same side and the aortic knuckle may be obscured by the collapse.

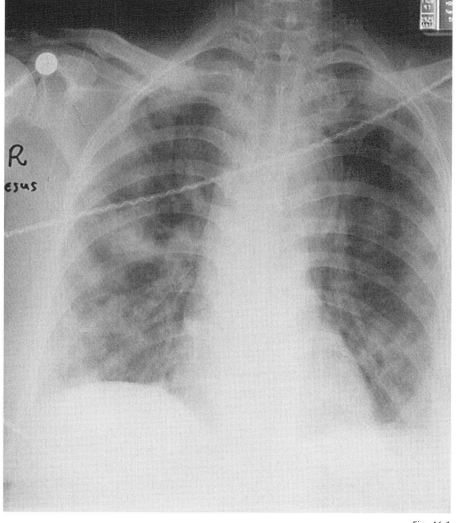

Fig. 46.1

Case 46. This was a 48-year-old male with fever of one week's duration. He was extremely ill and hypotensive requiring inotrope therapy. His CXR is shown (Fig. 46.1).

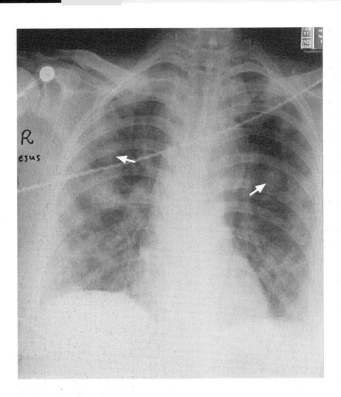

Fig. 46.2

CASE 46 DIFFUSE NODULAR INFILTRATES SUGGESTING BACTEREMIA AND SEPTIC LUNG ABSCESSES

The CXR shows nodules in both lungs (Fig. 46.2), which seem to be peripheral and of roughly equal size. The differential diagnosis would be cannon ball metastases though these are typically basal and of unequal size. This patient actually has Klebsiella bacteremia. In parts of South-East Asia, *Burkolderia pseudomallei* may result in the same CXR appearance. The other important etiologic agent is *Staphylococcus aureus* bacteremia.

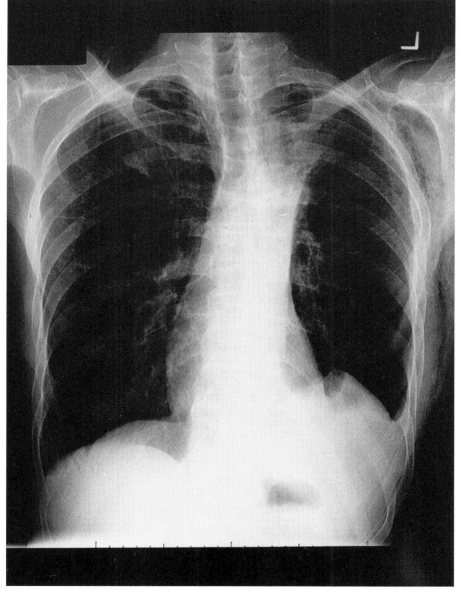

Fig. 47.1

Case 47. This patient gave a history of a recent left thoracotomy for massive hemoptysis. The CXR is shown (Fig. 47.1).

CASE 47 PREVIOUS LEFT UPPER LOBECTOMY

The CXR shows the left hemidiaphragm higher than the right, indicating volume loss of the left lung. Normally the left hemidiaphragm is about 1 cm lower than the right (at the height of the dome). The left main-stem bronchus is also more horizontal than usual, indicating volume loss in the left upper lobe. In addition, there is subcutaneous emphysema on the left chest wall.

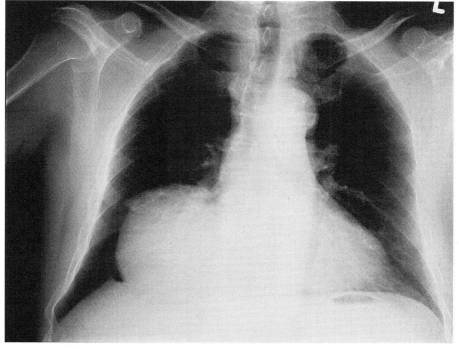

Fig. 48.1

Case 48. This patient was asymptomatic. His CXR is shown (Fig. 48.1).

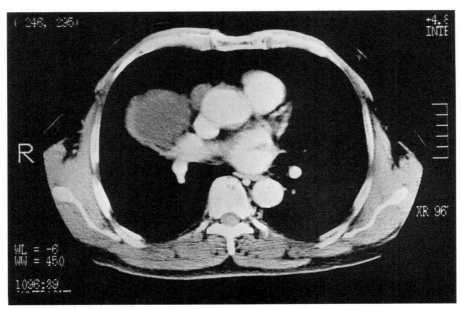

Fig. 48.2

CASE 48 PERICARDIAL CYST

The CXR shows a homogeneous opacity in the right cardio-phrenic angle. This opacity has a rounded border and sharp margins. The right heart border and the diaphragm are obliterated. CT (Fig. 48.2) shows the mass to be cystic with low-density material. All these are features of a pericardial cyst (also called spring water cyst).

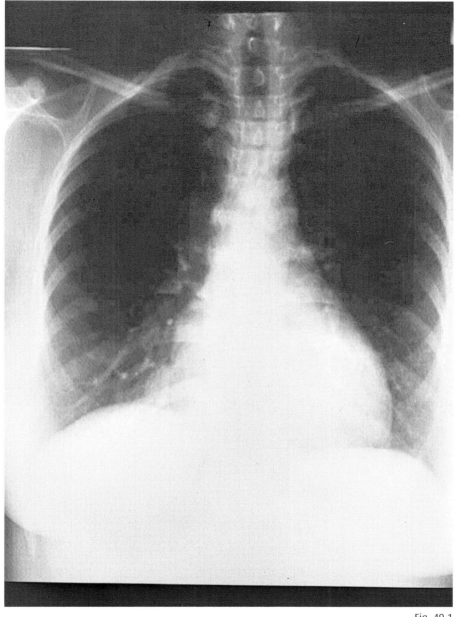

Fig. 49.1

Case 49. This patient was asymptomatic. Her CXR is shown (Fig. 49.1).

Fig. 49.2

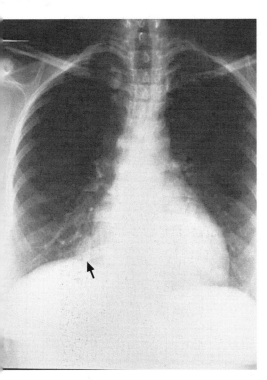

Fig. 49.3

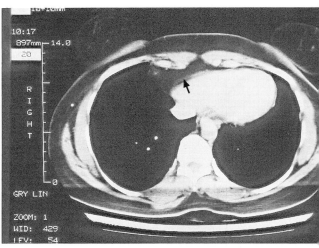

CASE 49 PERICARDIAL FAT PAD

The CXR shows opacity in the right cardio-phrenic angle (Fig. 49.2). Again, the opacity has obliterated the right heart margin and the diaphragm. CT (Fig. 49.3) shows the density to be fat making this a pericardial fat pad.

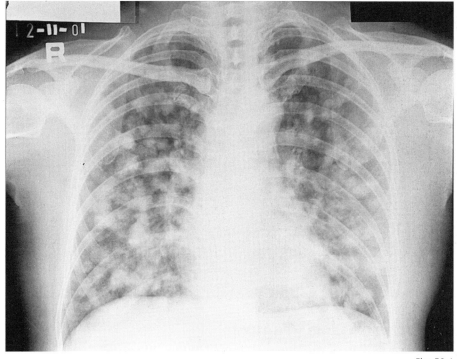

Fig. 50.1

Case 50. This patient with a history of carcinoma of the colon presented with chronic cough and loss of weight. The CXR is shown (Fig. 50.1).

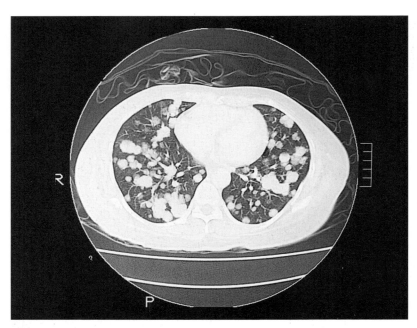

Fig. 50.2

CASE 50 METASTASES TO LUNGS

See Case 46. The CXR shows bilateral peripheral lung nodules of varying sizes and this is better demonstrated on the CT (Fig. 50.2). This appearance is typical of lung metastases. The basal predominance is due to the greater blood supply in the lung bases. Lung metastases can arise from cancers of the breast, colon, rectum, and kidney.

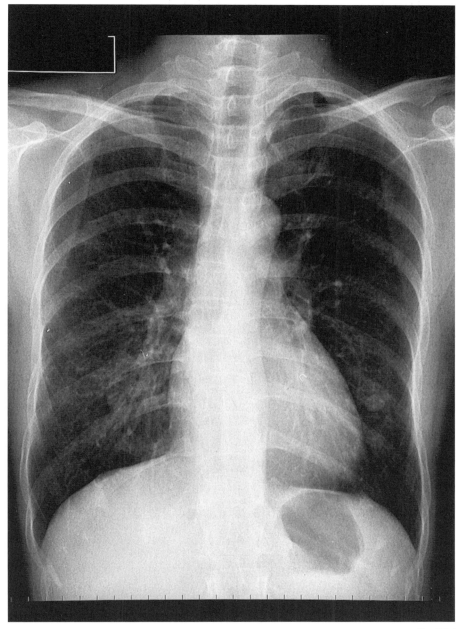

Fig. 51.1

Case 51. This female patient was asymptomatic. Her CXR is shown (Fig. 51.1).

Fig. 51.2

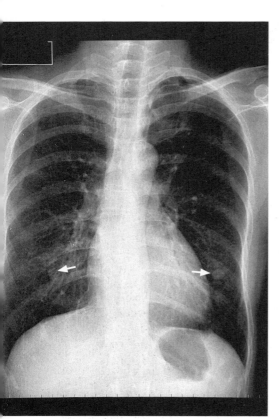

Fig. 51.3

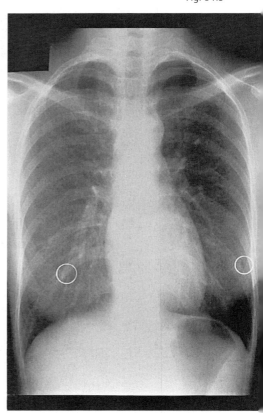

CASE 51 PULMONARY PSEUDO-NODULES DUE TO NIPPLE SHADOWS

There are two nodules (Fig. 51.2), one in each lower zone where the nipples are supposed to be. These shadows are typically homogeneous in appearance with sharp margins or sharp lateral margins and an absent medial margin. For patients with asymmetric nipples, the diagnosis can be difficult but a repeat CXR (Fig. 51.3) with nipple markers can help confirm that the opacity is due to a nipple.

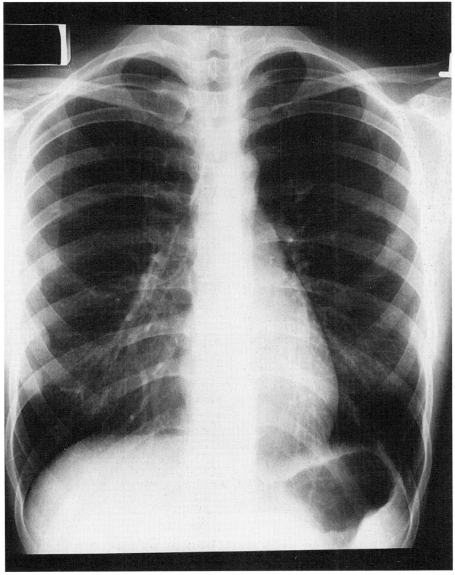

Fig. 52.1

Case 52. This young female had been coughing for the past few weeks. She also had right-sided pleuritic chest pain. Describe the most obvious CXR abnormality (Fig. 52.1).

Fig. 52.2

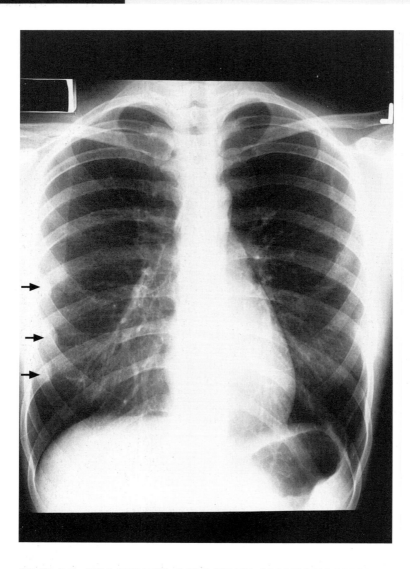

CASE 52   FRACTURED RIBS WITH CALLUS FORMATION

The CXR shows densities along the anterolateral aspect of the right fifth, sixth, and seventh ribs (Fig. 52.2). This appearance is consistent with callus formation along the ribs which could be due to cough fractures. An alternate way to view the rib fracture more clearly is a right lateral oblique film.

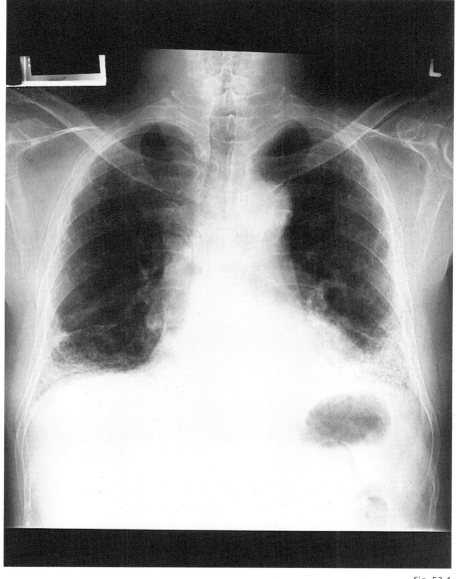

Fig. 53.1

Case 53. This middle-aged female presented with a one-year history of exertional dyspnea. Examination reveals clubbing, and chest auscultation revealed velcro-like crepitations. Her CXR is shown (Fig. 53.1).

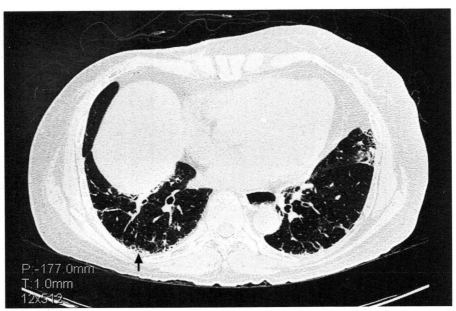

Fig. 53.2

CASE 53 IDIOPATHIC PULMONARY FIBROSIS

The CXR shows small bilateral lung volumes. There are basal infiltrates which are peripheral and cystic in appearance, not unlike a honeycomb. These changes are better demonstrated on the CT (Fig. 53.2). This is the typical appearance of Idiopathic Pulmonary Fibrosis (also known as Cryptogenic Fibrosing Alveolitis). The typical profile is a middle-aged female with shortness of breath over months. It can be associated with connective tissue diseases like rheumatoid arthritis and systemic lupus erythematosis.

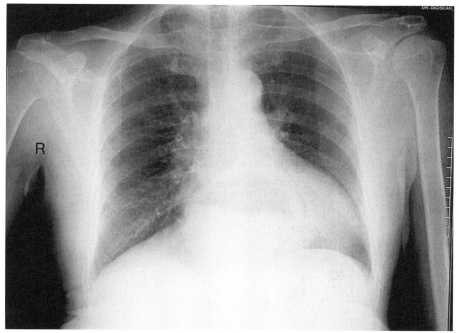

Fig. 54.1

Case 54. This middle-aged woman had symptoms of reflux. This was her CXR (Fig. 54.1).

CASE 54 HIATUS HERNIA

The CXR shows a lucent shadow with an air fluid level in the lower mediastinum. This is typical of a hiatus hernia because of its midline position with the stomach herniating through the esophageal hiatus. Also the stomach bubble is not seen in its usual position. A barium swallow or CT with oral contrast can be done in doubtful cases.

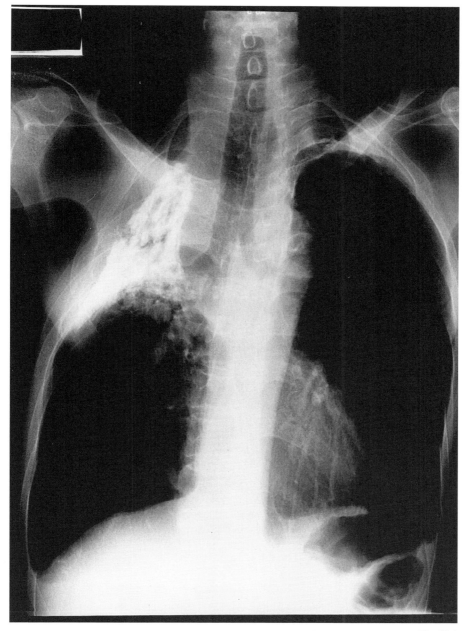

Fig. 55.1

Case 55. This patient gave a history of tuberculosis in the 1950s for which surgery was performed. The CXR is shown (Fig. 55.1).

CASE 55 PREVIOUS RIGHT THORACOPLASTY

Prior to the advent of effective anti-tuberculous drugs, surgery was the only treatment available for patients with tuberculosis. The objective was to cause closure of the upper lobe cavities and one option was thoracoplasty which involves resection of the upper ribs, resulting in lung collapse. In this CXR, the right upper chest is deformed and the pleural space is calcified. Other procedures performed include artificial pneumothorax, phrenic nerve crush, or plombage.

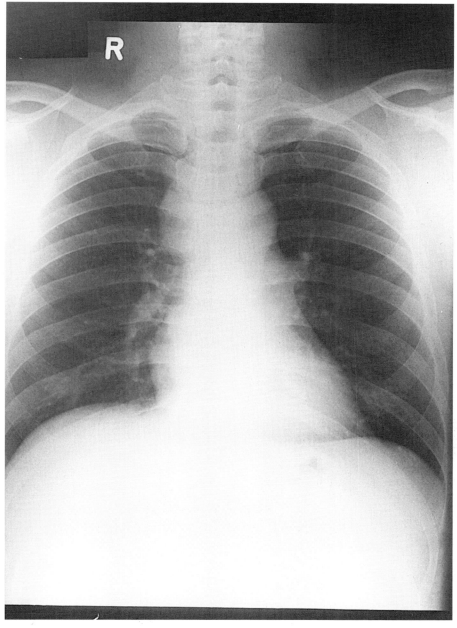

Fig. 56.1

Case 56. This middle-aged male was asymptomatic. His CXR (Fig. 56.1) remained unchanged for many years.

Fig. 56.2

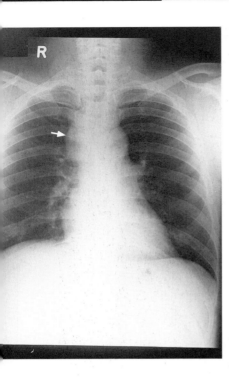

Fig. 56.3

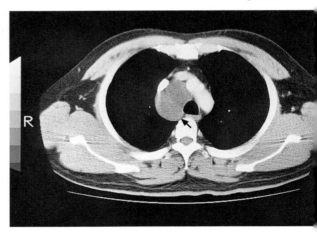

CASE 56 BRONCHOGENIC CYST

The CXR shows a bulge at the right paratracheal stripe (Fig. 56.2). The right para-tracheal stripe on an erect CXR is normally up to 10 mm wide. Other causes of a widened right paratracheal stripe include lymphoma, congestive cardiac failure, vascular abnormalities, and superior mediastinal masses. The CT (Fig. 56.3) shows a cystic (low-density) mass at the right paratracheal area, likely to be due to a congenital bronchogenic cyst. Bronchogenic cysts can occur in any part of the mediastinum but typical sites include the carina, paratracheal, retrocardiac areas, and adjacent to the esophagus in contact with the trachea or main bronchi.

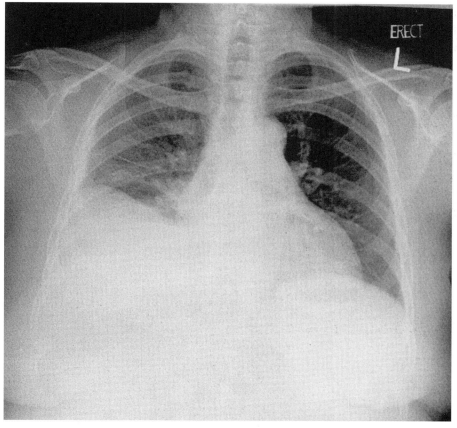

Fig. 57.1

Case 57. This patient gave a history of liver cirrhosis and ascites. The CXR is shown (Fig. 57).

CASE 57 RIGHT SUBPULMONIC EFFUSION

The CXR shows that the right costophrenic angle is blunted suggestive of a small pleural effusion. In addition, the right hemidiaphragm has its highest point displaced laterally. Normally the dome of the hemidiaphragm should have its highest point medial to the midpoint between the midline and the chest wall. These are clues to the fact that there is fluid trapped in the space between the right hemidiaphragm and the inferior aspect of the lung.

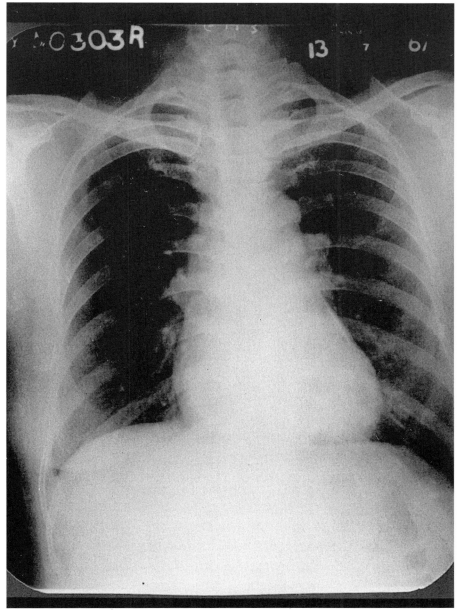

Fig. 58.1

**Case 58. This elderly patient is asymptomatic. He gave a history of a
prolonged severe viral illness previously. This is his CXR (Fig. 58.1).**

Fig. 58.2

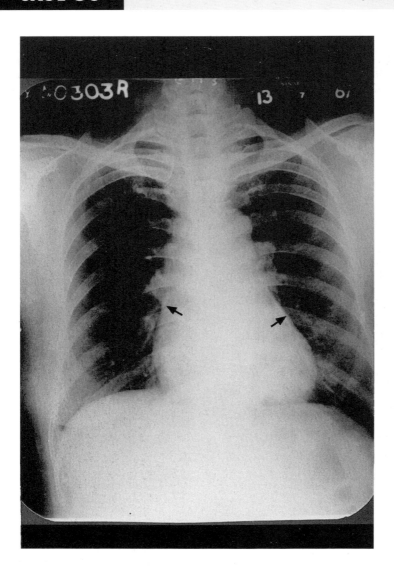

CASE 58 CHRONIC CALCIFIC PERICARDITIS

The CXR shows calcification of the pericardium (Fig. 58.2) indicative of previous chronic pericarditis. Causes include previous viral pericarditis, asbestos exposure, granulomatous disease like tuberculosis or histoplasmosis, mediastinal irradiation, or trauma. However, a significant number of such cases have no apparent cause.

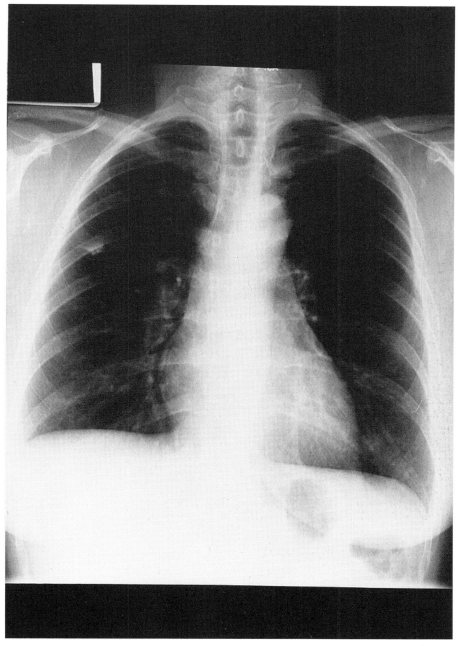

Fig. 59.1

Case 59. This patient was asymptomatic. The CXR is shown (Fig. 59.1).

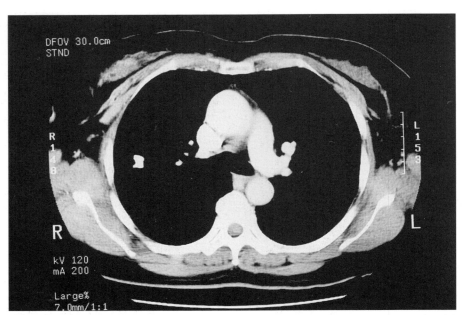

Fig. 59.2

CASE 59 RIGHT UPPER LOBE SPN DUE TO A CALCIFIED GRANULOMA

The CXR shows a dense right upper lobe solitary pulmonary nodule. The nodule is less than 1 cm (see Case 35) diameter and CT confirms it to be dense and homogeneously calcified (Fig. 59.2), a characteristic of previous granulomatous disease like histoplasmosis or tuberculosis.

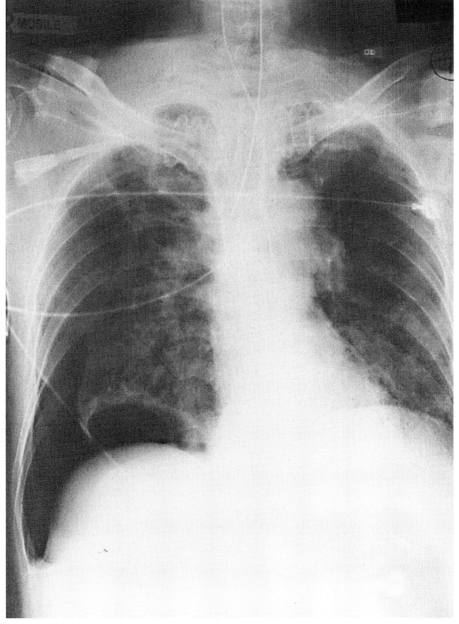

Fig. 60.1

Case 60. This patient was admitted to the ICU for septic shock requiring mechanical ventilation and inotropic support. This CXR was taken after admission (Fig. 60.1).

Fig. 60.2

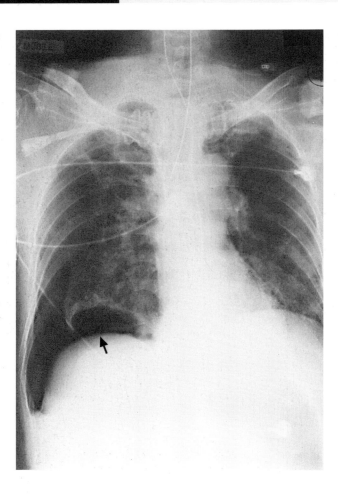

CASE 60 RIGHT TENSION PNEUMOTHORAX

See Case 2. The CXR shows that the endotracheal tube is too far down and the tip is now sitting at the origin of the right main-stem bronchus. The ideal position is for the tip of the tube to be at the level of the clavicles. This patient also had a right central venous catheter inserted. The tip of the central venous catheter should ideally lie at the junction of the superior vena cava and the right atrium. The other important finding is that of a lucent area at the anterior costophrenic recess on the right side with no lung markings. This is the deep sulcus sign and is indicative of a right pneumothorax (Fig. 60.2). In addition, the right hemidiaphragm is depressed and the mediastinum shifted away indicating a tension pneumothorax.

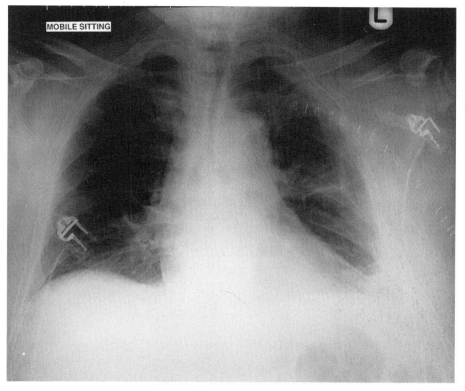

Fig. 61.1

Case 61. This middle-aged male was diagnosed as having asthma but has not improved following inhaled steroids. His CXR is shown (Fig. 61.1).

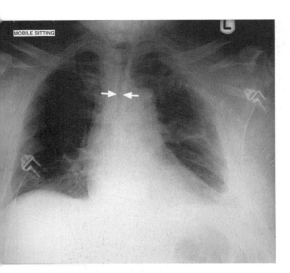

Fig. 61.2

Fig. 61.3

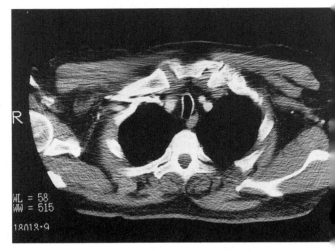

CASE 61   TRACHEAL STENOSIS DUE TO TRACHEOPATHIA OSTEOCHONDROPLASTICA

The CXR shows narrowing of the tracheal air column (Fig. 61.2) with calcification of the wall. Tracheal narrowing can be due to malignant causes (lung cancer, lymphoma, metastases) or benign causes (post tuberculosis, posttraumatic, amyloidosis, sarcoidosis, Wegener's, Tracheopathia Osteochondroplastica). Tracheopathia Osteochondroplastica (TO) is an extremely rare condition, characterized by the presence of multiple osseous and/or cartilaginous submucosal nodules (Fig. 61.3) protruding into the lumen of the airway. Bronchoscopy is diagnostic but treatment is nonspecific and supportive.

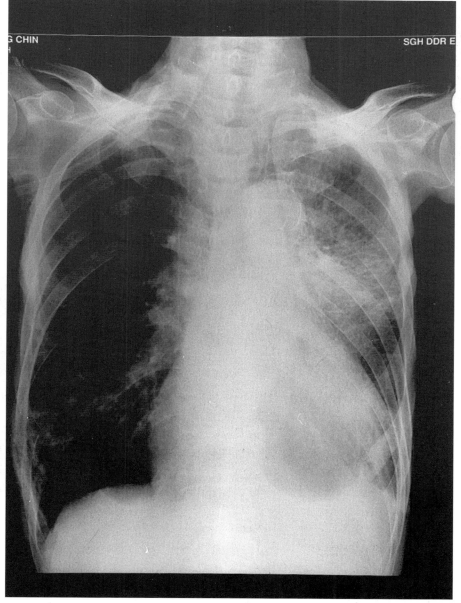

Fig. 62.1

Case 62. This middle-aged male gave a history of lung cancer. Recently, he complained of loss of weight and shortness of breath. His CXR is shown (Fig. 62.1).

Fig. 62.2

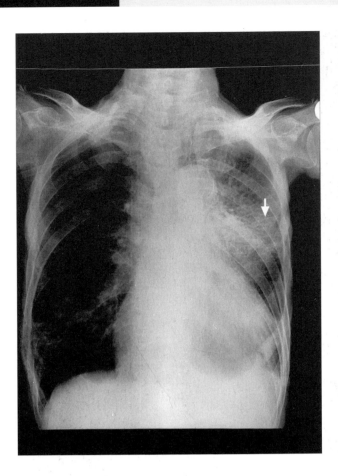

CASE 62 LUNG CANCER WITH LYMPHANGITIS CARCINOMATOSIS

The CXR shows a left upper lobe mass and mid-zone infiltrates with a normal heart size. In addition, there are Kerley B lines (Fig. 62.2) in the periphery of the left mid zone, suggestive of lymphatic distension. These features are consistent with the advanced lung cancer metastasizing to the lymphatics. The prognosis is extremely poor.

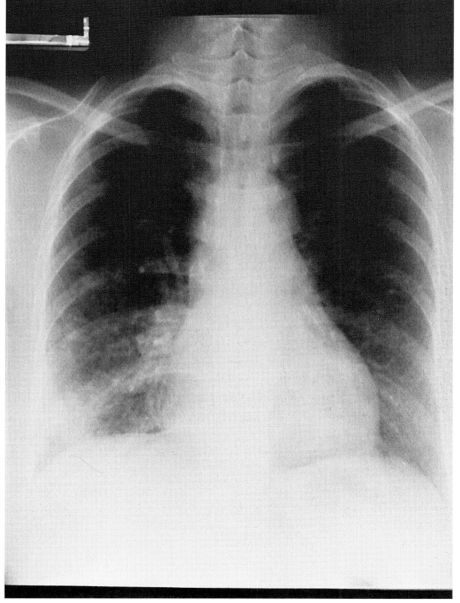

Fig. 63.1

Case 63. This 24-year-old female was asymptomatic. Six months ago, she presented with pneumonia-like symptoms of cough, fever, and purulent sputum. Describe her CXR (Fig. 63.1).

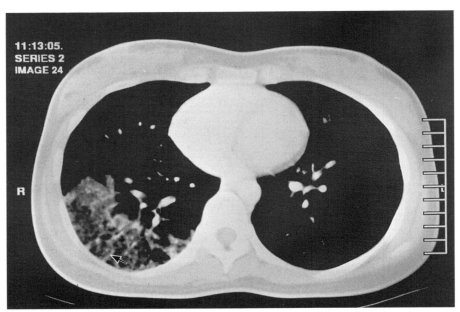

Fig. 63.2

CASE 63 BRONCHIOLITIS OBLITERANS ORGANIZING PNEUMONIA (BOOP)

The CXR shows a right lower lobe infiltrate which demonstrates some air bronchograms on CT (Fig. 63.2). In addition, there seems to be a beady appearance to the infiltrates. Pneumonic changes on CXR typically resolve within three months. She subsequently underwent a bronchoscopy and transbronchial lung biopsy which showed BOOP. This is an idiosyncratic reaction sometimes seen in association with drugs, chemical inhalation, connective tissue disease, and various infections. This is usually very steroid-responsive.

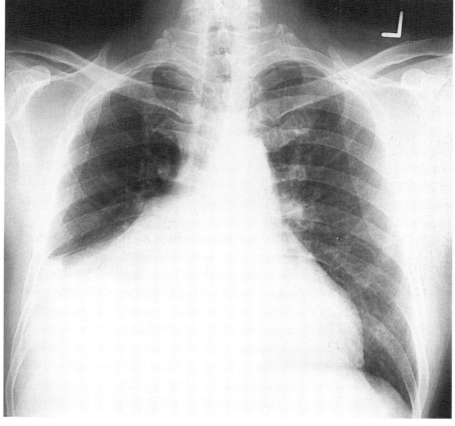

Fig. 64.1

Case 64. This elderly male had hemoptysis and loss of weight over the past three months. His CXR is shown (Fig. 64.1).

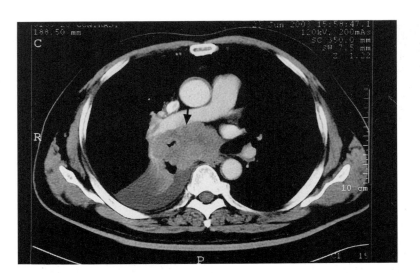

Fig. 64.2

CASE 64 MASS IN THE BRONCHUS INTERMEDIUS WITH COLLAPSE OF THE MIDDLE AND LOWER LOBE

The CXR shows a density in the right middle zone. The density is demarcated superiorly by a horizontal line, the transverse fissure, which is pulled down. The medial border of the mass has merged with the right heart border indicative of right middle lobe disease. The right hemidiaphragm is also obscured, indicating right lower lobe disease. These features are consistent with a mass arising from the bronchus intermedius with resultant collapse of the right middle and lower lobes (Fig. 64.2).

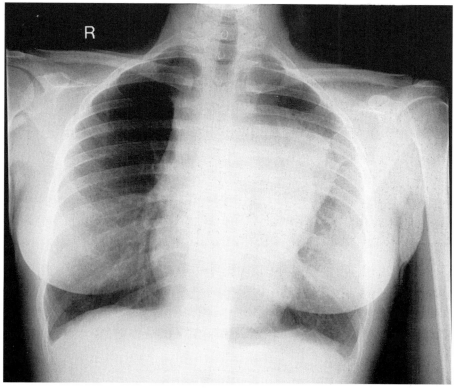

Fig. 65.1

Case 65. This young female had been breathless over the last two months. Her symptoms are worse on lying down. Her CXR (Fig. 65.1) is shown.

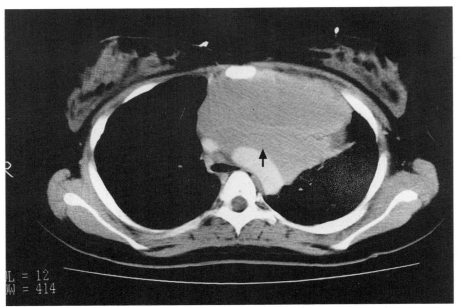

Fig. 65.2

CASE 65 ANTERIOR MEDIASTINAL MASS DUE TO LYMPHOMA

The PA CXR shows a mass adjacent to the left heart border. There is hyperinflation of both lung fields, suggesting obstructive airway disease. The left cardiac margin is obscured, indicating an anterior mediastinal mass as the heart is an anterior mediastinal structure. CT (Fig. 65.2) confirms that there is a mass in the anterior mediastinum and this mass is compressing the lower trachea and main-stem bronchi and right pulmonary artery. The differential diagnoses of masses in the anterior mediastinum include the 5 "T"s: thyroid masses, teratoma, thymic masses, (terrible) lymphoma, and thoracic aneurysm.

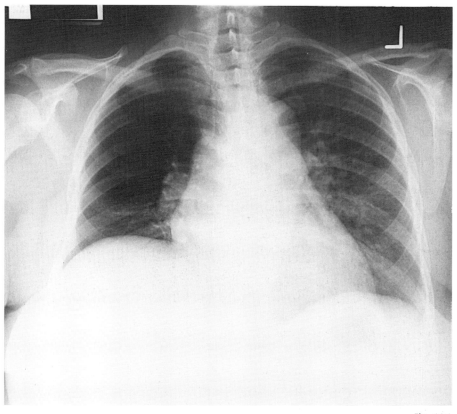

Fig. 66.1

Case 66. This 47-year-old female had streaky hemoptysis for two years associated with dyspnea on exertion. What does the CXR show (Fig. 66.1)?

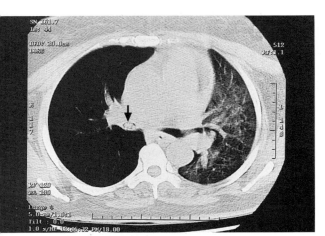

Fig. 66.2

Fig. 66.3

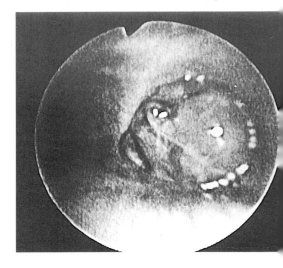

CASE 66 RIGHT LUNG OLIGEMIA – CARCINOID TUMOR OF RIGHT MAIN STEM BRONCHUS

The CXR shows a hyper-lucent right lung associated with volume loss as indicated by an elevated right hemidiaphragm. CT confirms the presence of the mass in the right main-stem bronchus and the oligemic right lung (Fig. 66.2). Air trapping may be demonstrated on an expiratory CXR showing an exaggeration of the oligemia and the shifting away of the mediastinum. In this patient, bronchoscopy showed a slow-growing carcinoid tumor in the right main-stem bronchus (Fig. 66.3). The differential diagnoses of a hyper-lucent lung are bullae, acute pulmonary embolism, pneumothorax, Macleod's syndrome, and a ball-valve-effect type of bronchial obstruction.

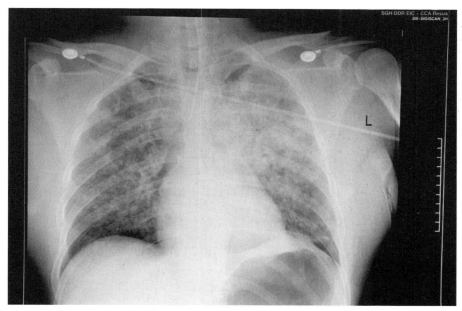

Fig. 67.1

Case 67. This patient presented with stridor due to thyroid goiter and was intubated (Fig. 67.1). Repeat CXR was done six hours later (Fig. 67.2). What is the main radiological abnormality? What is the cause?

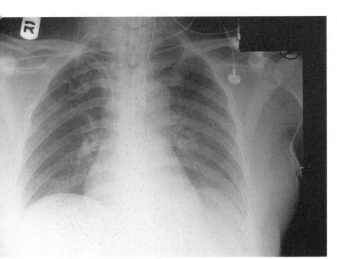

Fig. 67.2

Fig. 67.3

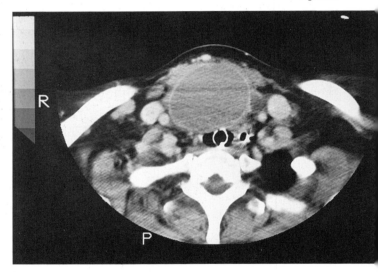

CASE 67 FLASH PULMONARY EDEMA DUE TO UPPER AIRWAY OBSTRUCTION

The first CXR shows a normal cardiac shadow associated with bilateral perihilar alveolar infiltrates suggestive of acute pulmonary edema. The development of pulmonary edema with a normal heart size is indicative of an acute event. The rapid clearance of the pulmonary infiltrates here indicates that the process is rapidly corrected by positive pressure. In this patient, an important consideration is negative pressure pulmonary edema due to upper airway obstruction from the thyroid goiter, which is seen on the CT (Fig. 67.3).

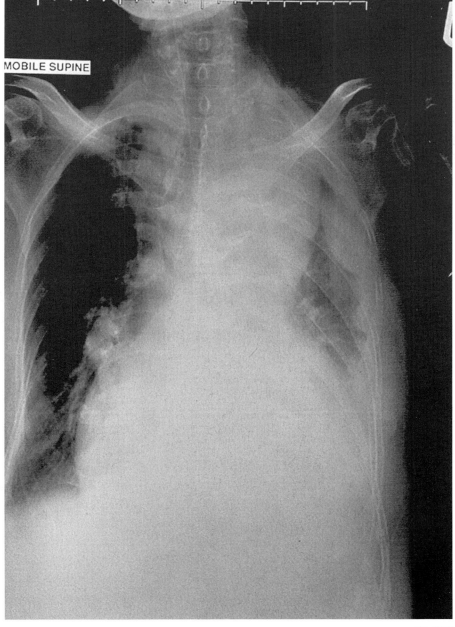

MOBILE SUPINE

Fig. 68.1

Case 68. This elderly female presented with left-sided chest pain of three months' duration. Name the CXR abnormalities (Fig. 68.1).

Fig. 68.2

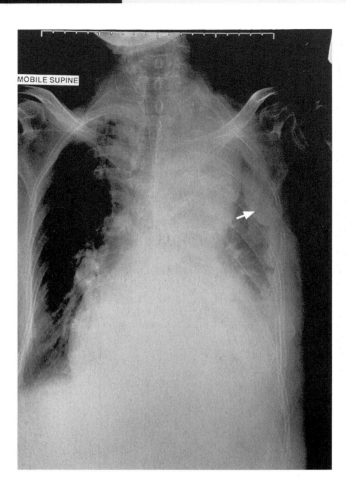

CASE 68 LEFT PLEURAL EFFUSION AND LYTIC LESION IN THE LEFT THIRD RIB SUGGESTIVE OF METASTATIC DISEASE

The CXR shows a moderate-sized left pleural effusion, which is loculated. There is also globular cardiomegaly, suggesting a pericardial effusion. Pleural tap showed malignant cells consistent with the diagnosis of adenocarcinoma of the lung. In addition, the second, third, and fourth ribs on the left side (Fig. 68.2) demonstrate lytic lesions in keeping with bony metastases. Bone scan would be helpful in confirming the bone metastases. These are all features of advanced lung cancer with metastatic involvement.

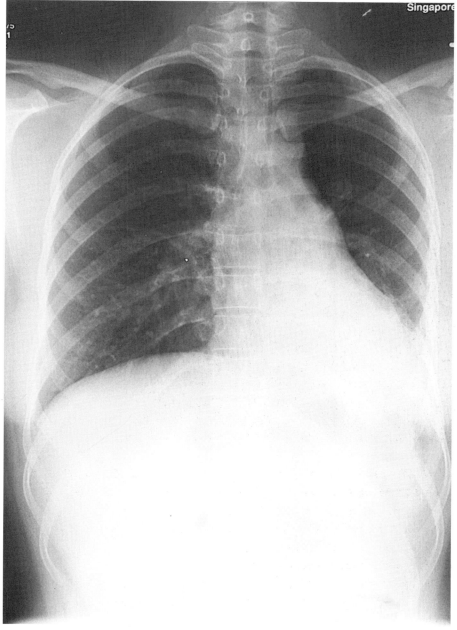

Fig. 69.1

Case 69. This 50-year-old female with a past history of tuberculosis had chronic cough over the past year. Describe her CXR (Fig. 69.1).

Fig. 69.2

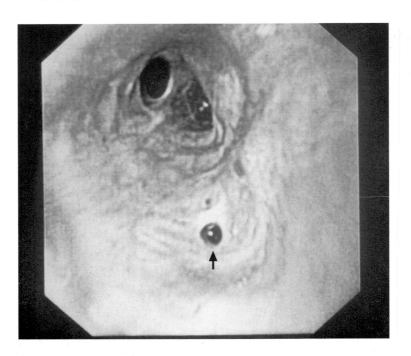

CASE 69 LEFT LOWER LOBE COLLAPSE

There is volume loss in the left lung as indicated by an elevation of the left hemidiaphragm and shift of mediastinum to the left. The left hemithorax is also smaller
than the right. In addition, the left hemidiaphragm is obscured indicating a left
lower lobe collapse. At bronchoscopy, she was found to have a benign stricture of
the left lower lobe orifice (Fig. 69.2) from previous tuberculosis.

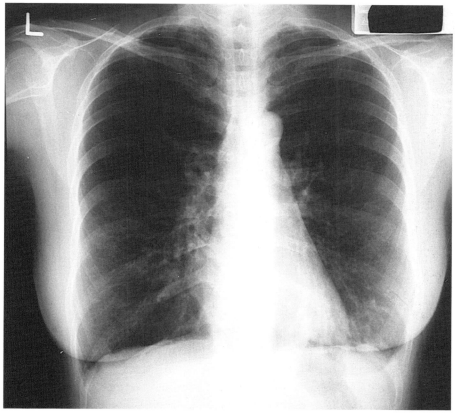

Fig. 70.1

Case 70. This 35-year-old female had a long history of chronic productive cough. Her CXR is shown (Fig. 70.1).

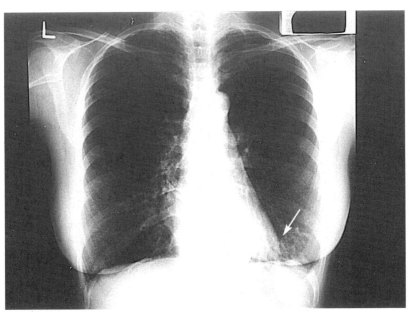

Fig. 70.2

CASE 70 DEXTROCARDIA DUE TO KARTAGENER'S SYNDROME

This patient has obvious dextrocardia (the heart is on the right side) and situs inversus (the stomach bubble is also on the right side instead of the left). There is also right lower lobe bronchiectasis (Fig. 70.2) as evidenced by bronchial wall thickening, bronchial opacification (bronchocele), and loss of volume. Dextrocardia and situs inversus may be associated with ciliary dysfunction causing sinusitis and bronchiectasis. This is called Kartagener's Syndrome.

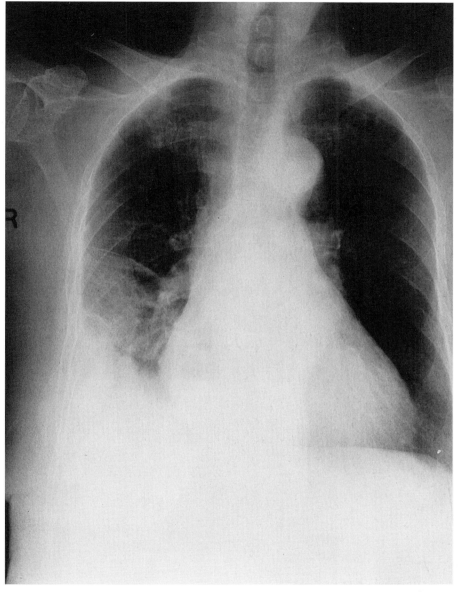

Fig. 71.1

Case 71. This elderly male alcoholic had a binge and subsequently pre-sented with alcoholic intoxication and vomiting. His CXR is shown (Fig. 71.1). What is the main abnormality?

Fig. 71.2

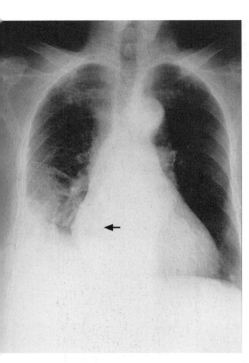

Fig. 71.3

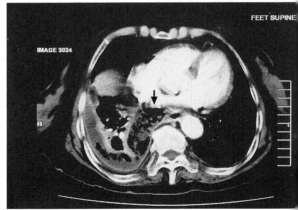

CASE 71 MEDIASTINITIS

There is a right-sided pleural effusion and, in addition, an air-fluid level is noted behind the right side cf the heart (Fig. 71.2). This is typical of a perforated esophagus (Boerhaave's Syndrome due to a full thickness laceration leading to mediastinitis from vomiting) resulting in free air in the mediastinum and a pleural effusion. This patient had food particles at tube thoracostomy. CT scan shows the right hydropneumothorax due to the resultant empyema (Fig. 71.3).

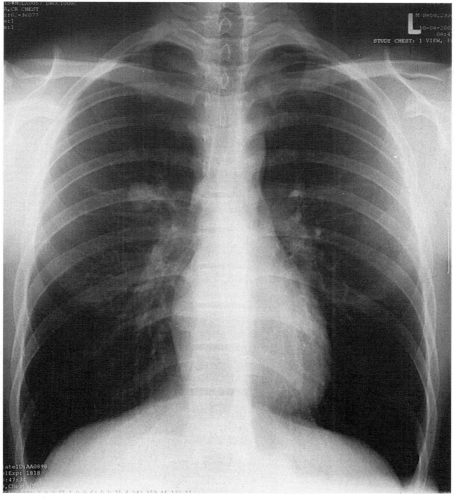

Fig. 72.1

Case 72. This elderly male was totally asymptomatic (Fig. 72.1). What does the CXR show?

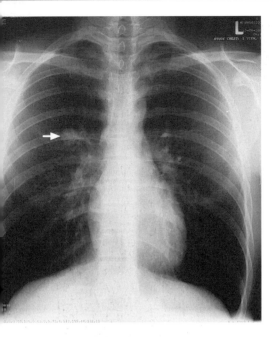

Fig. 72.2

Fig. 72.3

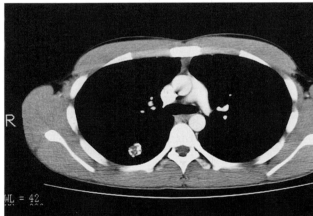

CASE 72 SPN DUE TO HAMARTOMA

The CXR (Fig. 72.2) shows a very well-demarcated and dense-looking SPN in the
right upper lobe. The CT (Fig. 72.3) confirms the nodule to be calcified with a
popcorn pattern, typical of a hamartoma. A hamartoma is a congenital overgrowth
of mature cells and is a common cause of a benign pulmonary nodule. They are
typically well demarcated and usually contain fat and/or calcium.

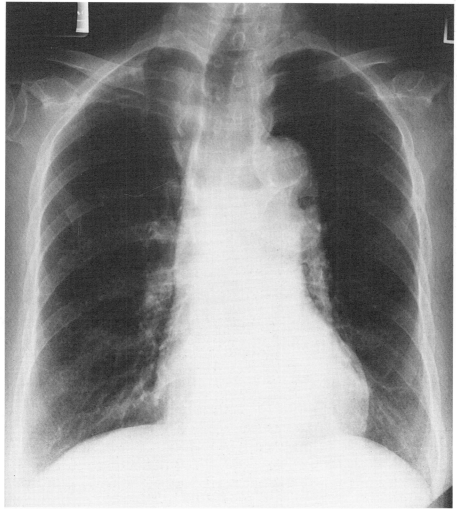

Fig. 73.1

Case 73. This elderly female gave a history of dysphagia and loss of weight of two months' duration. Her CXR is shown (Fig. 73.1). What is the diagnosis?

Fig. 73.2

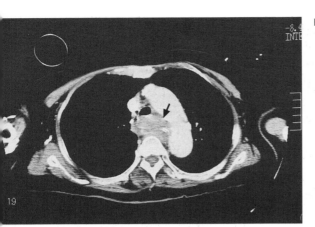

Fig. 73.3

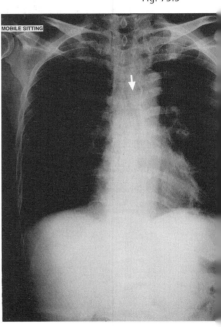

CASE 73 CARCINOMA OF ESOPHAGUS

The CXR shows a dilated esophagus behind the heart with an air-fluid level in the region of the carina. This is indicative of an obstruction at the esophagus and possibilities include carcinoma of the esophagus, benign stricture, achalasia, and scleroderma. The CT Thorax (Fig. 73.2) shows the esophageal carcinoma more clearly at the level of the carina. Fig. 73.3 shows a similar patient who has had an esophageal stent (Ultraflex) placed for palliation of dysphagia.

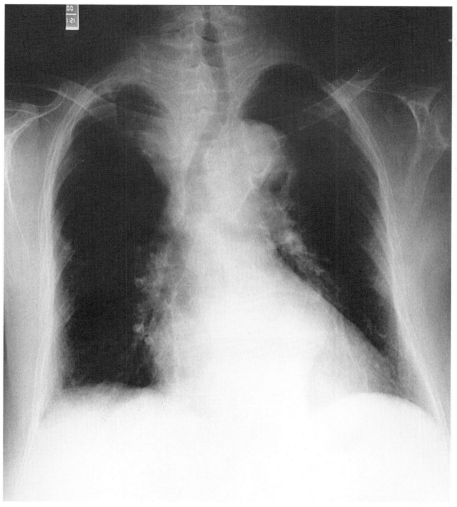

Fig. 74.1

Case 74. This 70-year-old female was asymptomatic. Her CXR (Fig. 74.1) is shown.

Fig. 74.2

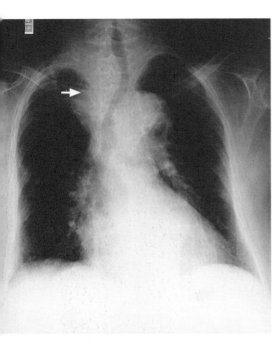

Fig. 74.3

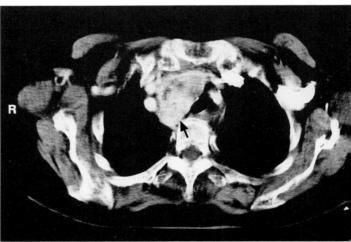

CASE 74 RIGHT UPPER ZONE OPACITY DUE TO RETROSTERNAL GOITER

The CXR shows a shadow in the right upper zone distending the right paratracheal stripe. The shadow has a sharp curvilinear lateral border (Fig. 74.2) and seems to extend to the neck. The trachea is also displaced to the left. Such a shadow at this location is typical of retrosternal goiter and this is confirmed on the CT (Fig. 74.3).

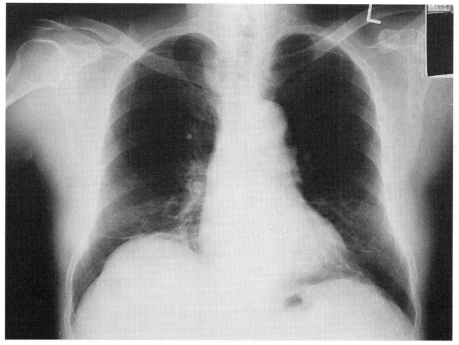

Fig. 75.1

Case 75. This patient was asymptomatic. His CXR is shown (Fig. 75.1).

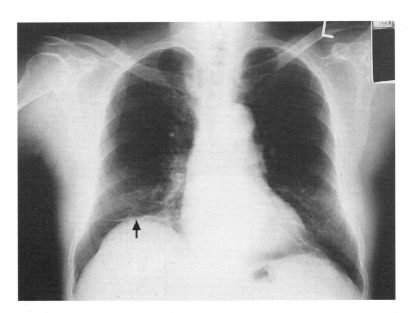

Fig. 75.2

CASE 75 EVENTRATION OF RIGHT HEMIDIAPHRAGM

The CXR shows that the right hemidiaphragm has a hump (Fig. 75.2) causing it to appear as if it is much more elevated than the left hemidiaphragm. This is usually due to a congenital weakness of the tendinous part of the hemidiaphragm and is of no clinical significance. Normally, the right hemidiaphragm is about 1–2 cm higher than the left.

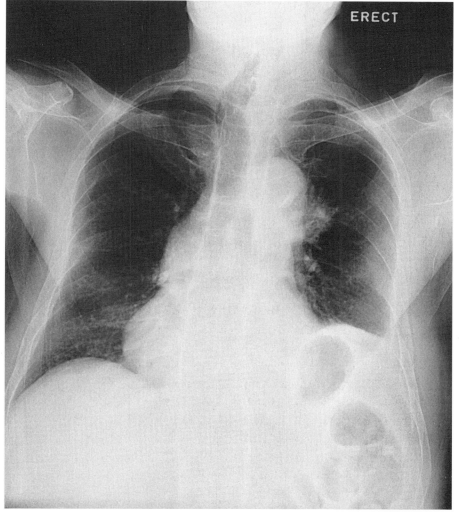

Fig. 76.1

Case 76. This elderly female smoker presented with loss of weight. Her CXR is shown (Fig. 76.1).

Fig. 76.2

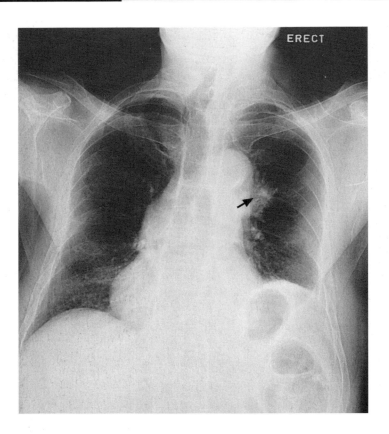

ERECT

CASE 76 MASS IN LEFT LUNG WITH ELEVATED LEFT HEMIDIAPHRAGM

The CXR shows a mass in the left upper lobe (Fig. 76.2), representing a primary lung cancer. In addition, the left hemidiaphragm is elevated such that it is higher than the right. Normally the right hemidiaphragm is 1–2 cm higher than the left. The left hemidiaphragm elevation here is likely due to left phrenic nerve palsy, probably due to mediastinal lymph node metastases. Other causes of an elevated hemidiaphragm include recent cardiac bypass surgery, trauma, and previous herpes zoster involving the phrenic nerve. The commonest cause, however, is idiopathic phrenic nerve palsy.

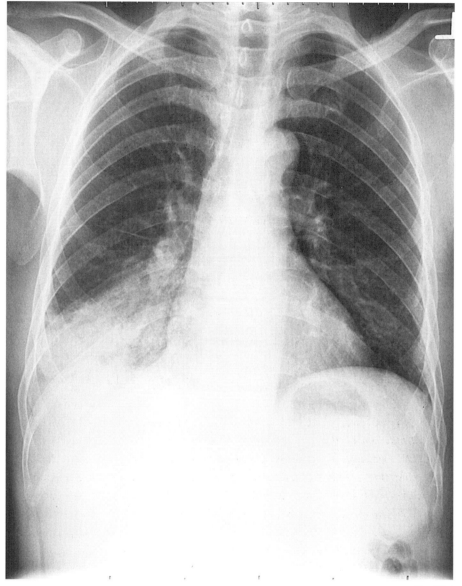

Fig. 77.1

Case 77. This middle-aged male had low-grade fever of one month's duration associated with productive cough and loss of weight. His CXR is shown (Fig. 77.1).

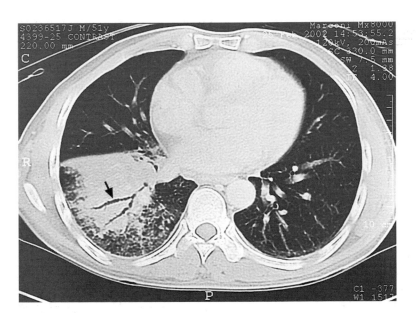

Fig. 77.2

CASE 77 RIGHT LOWER LOBE CONSOLIDATION DUE TO TUBERCULOSIS IN HUMAN IMMUNODEFICIENCY VIRUS (HIV) HOST

The patient had subacute fever and the CXR shows air bronchograms in the right lower zone. There is loss of outline of the right hemidiaphragm (Silhouette sign) confirming right lower lobe consolidation (Fig. 77.2). His sputum turned out to be positive on acid-fast bacillus (AFB) smear confirming that he had active pulmonary tuberculosis. He also reported frequent visits to commercial sex workers and his HIV serology was positive. There are two forms of TB in HIV patients. One form, occurring in early HIV disease, is no different from that in a non-immuno-compromised host with classic upper lobe cavitary disease. In patients with late stage HIV disease, CXR presentation is atypical with less preponderance of cavitation, less upper lobe disease, and greater predominance of thoracic lymphadenopathy and pleural effusion. All patients with tuberculosis should be investigated for HIV infection.

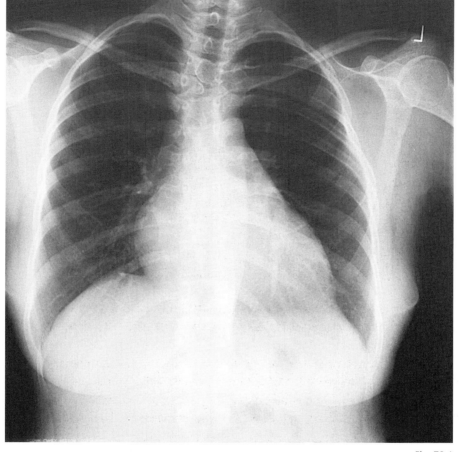

Fig. 78.1

Case 78. This patient was asymptomatic (Fig. 78.1). The CXR is shown.

Fig. 78.2

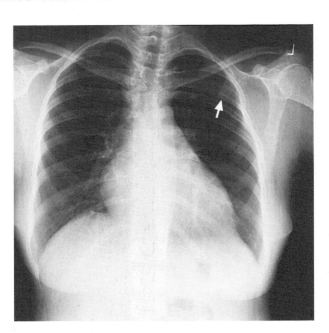

CASE 78 FIBROUS DYSPLASIA INVOLVING LEFT FOURTH RIB

The fourth rib on the left side is expanded with evidence of thinning of the cortex (Fig. 78.2). The same rib is bifid anteriorly. The ribs are the most common sites of solitary fibrous dysplasia with the most common location in the lateral or posterior portion of the ribs. In this condition, the bone is replaced with fibrous tissue. Bone scan is usually negative.

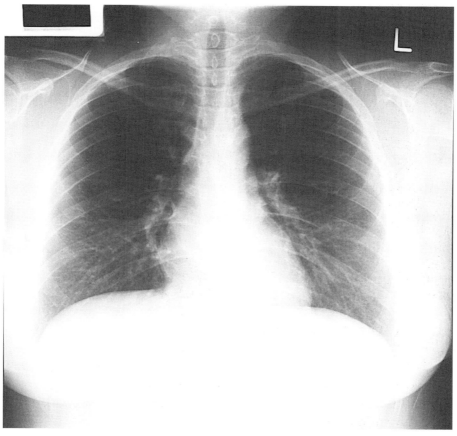

Fig. 79.1

Case 79. This patient was asymptomatic (Fig. 79.1).

Fig. 79.2

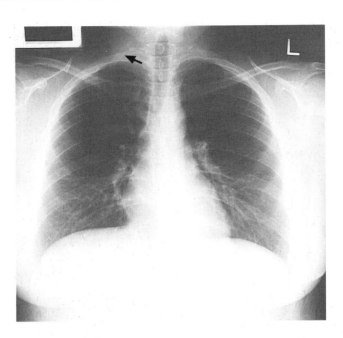

CASE 79 RIGHT CERVICAL RIB

The CXR shows a right accessory rib, which is arising from the seventh cervical vertebra (Fig. 79.2). This condition is present in 1% of the population with 80% of cases being bilateral. Rarely, this condition may be symptomatic due to pressure on the lower trunk of the brachial plexus.

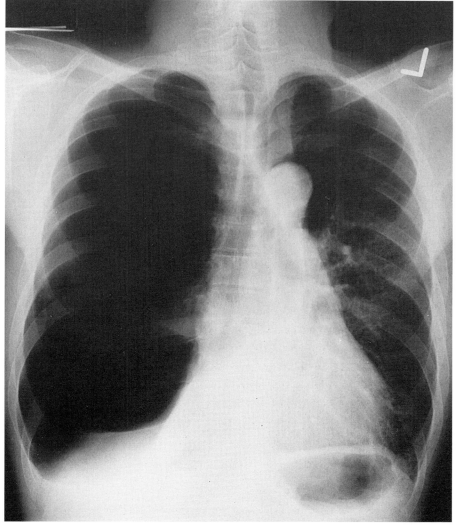

Fig. 80.1

Case 80. This elderly male chronic smoker had a five-year history of pro-ductive cough and exertional dyspnea. His CXR is shown (Fig. 80.1).

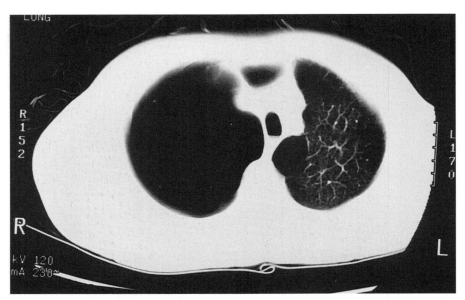

Fig. 80.2

CASE 80 COPD WITH BULLAE

In addition to features of hyperinflation (Case 19), the CXR also shows right lung hyperlucency. This could be due to two possibilities, pulmonary embolus or bulla. Right pneumothorax is unlikely because of the absence of a pleural line. CT (Fig. 80.2) demonstrates the presence of a giant right upper lobe bulla, a complication of COPD. One must be careful to make this differentiation, as tube thoracostomy in a patient with a bulla can be disastrous.

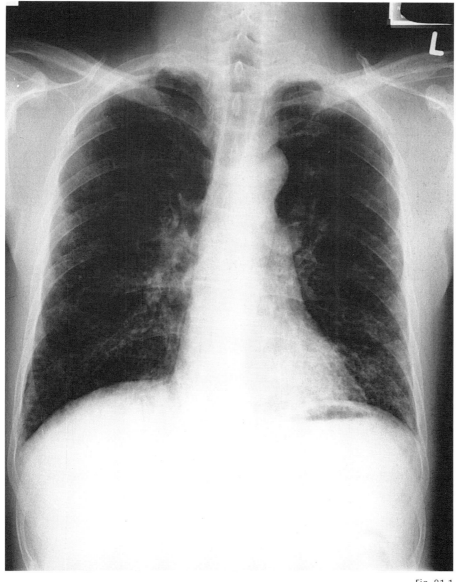

Fig. 81.1

Case 81. This middle-aged male of Japanese origin had chronic rhinitis and productive cough for a few years. The CXR is shown (Fig. 81.1).

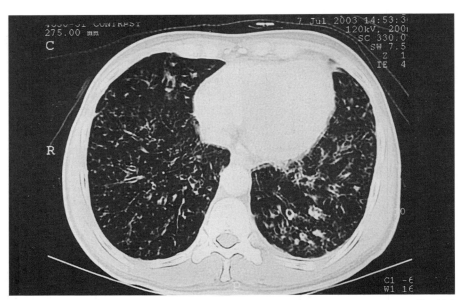

Fig. 81.2

CASE 81 BILATERAL SMALL RING SHADOWS – DIFFUSE PANBRONCHIOLITIS

The CXR shows bilateral diffuse shadows, which appear ring-like with tram-lining of the airways (best seen behind the heart). In some areas they may appear nodular. The long duration makes metastatic cancer unlikely. High resolution CT Thorax (Fig. 81.2) shows thickened bronchi and bronchioles. The small peripheral airways have a "tree-in-bud" appearance (due to mucoid impaction of the small airways) consistent with the diagnosis of diffuse panbronchiolitis. This condition was first described by the Japanese and is associated with an obstructive pulmonary physiology on testing. Treatment consists of a long course of low-dose macrolide antibiotics.

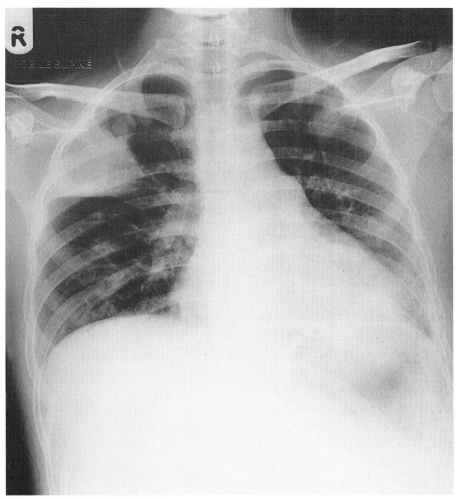

Fig. 82.1

Case 82. A young male with acute myeloid leukemia underwent a bone marrow transplant. This was complicated by relapse of the leukemia and persistent neutropenic fever. The CXR is shown (Fig. 82.1). A CXR two months ago was normal.

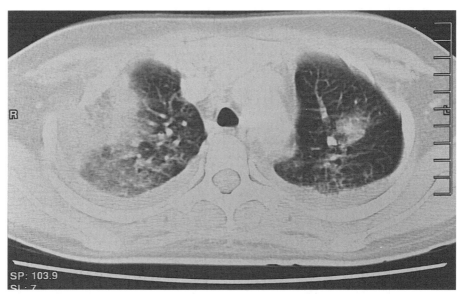

Fig. 82.2

CASE 82 BILATERAL UPPER LOBE NODULES DUE TO ANGIO-INVASIVE PULMONARY ASPERGILLOSIS

One of the most feared complications of severe neutropenic (<500 neutrophils per mm^3) sepsis in the immunosuppressed patient is invasive pulmonary aspergillosis. The CXR shows a right upper lobe mass, which is wedge shaped with the apex towards the hilum. This shadow is suggestive of a pulmonary infarct. In addition, the left upper lobe shows a small nodule at the periphery. The fact that CXR recently was normal makes a severe overwhelming infection very likely. The CT (Fig. 82.2) demonstrated two additional findings. The right upper lobe mass has a necrotic center and a surrounding halo. This is the classic "halo sign" (ground glass change adjacent to central dense consolidation) around the right upper lobe mass. The halo is thought to represent edema or hemorrhage due to infection by angiotrophic organisms, the most common being *Aspergillus fumigatus*.

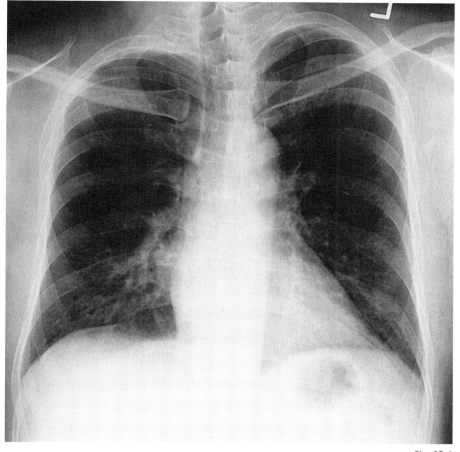

Fig. 83.1

Case 83. This patient had recurrent hemoptysis. The CXR is shown (Fig. 83.1).

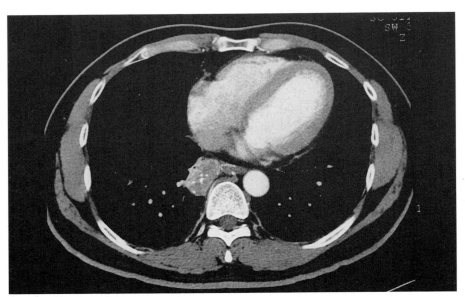

Fig. 83.2

CASE 83 MASS BEHIND RIGHT HEART BORDER DUE TO SEQUESTRATED LUNG

There is a mass seen along the right heart border. The fact that the margin of the heart is not silhouetted out implies that the mass is likely to be in the posterior mediastinum (the heart is an anterior structure), making lung cancer unlikely. CT (Fig. 83.2) confirmed the presence of the mass and, in addition, blood vessels are seen arising from the descending aorta making the diagnosis of sequestrated lung likely. This is a congenital abnormality where the non-functioning lung tissue is typically not in continuity with the tracheobronchial tree and derives its blood supply from systemic vessels. The normal lung encloses intralobar sequestration whereas extralobar sequestration is enclosed by its own pleura.

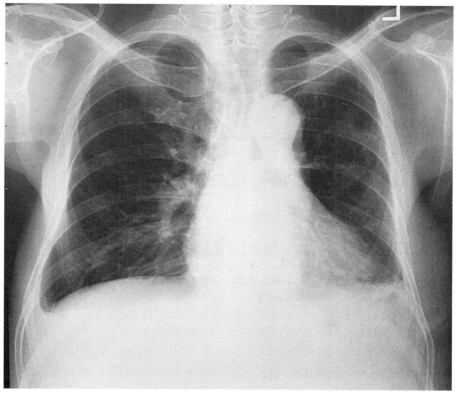

Fig. 84.1

Case 84. This 60-year-old male presented with dyspnea on exertion, abdominal distension, and loss of appetite and weight. His CXR is shown (Fig. 84.1).

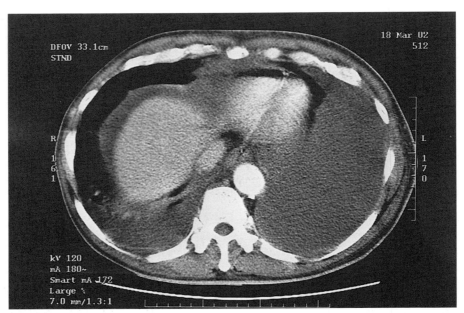

Fig. 84.2

CASE 84 BILATERAL PLEURAL EFFUSIONS DUE TO INTRA-ABDOMINAL MALIGNANCY AND CARCINOMATOUS PERITONEII

The CXR shows both costophrenic angles being blunted suggesting bilateral pleural effusions. The heart size is normal. In situations of bilateral pleural effusion, one must think in terms of a transudative state, e.g. congestive heart failure, nephrotic syndrome, or liver cirrhosis. If the fluid is exudative, an intra-abdominal process has to be ruled out. In the CXR shown, another important finding is the small lung volumes, which may suggest an intra-abdominal process, e.g. malignant ascites tracking through the diaphragmatic foramina (Fig. 84.2).

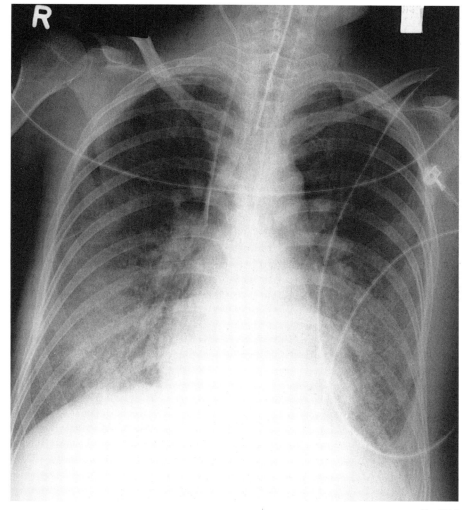

Fig. 85.1

Case 85. This patient was admitted for severe acute pancreatitis. A few days into the admission, the patient became very tachypneic and required intubation and mechanical ventilation. This is the CXR (Fig. 85.1).

CASE 85 ACUTE RESPIRATORY DISTRESS SYNDROME (ARDS)

The CXR shows bilateral diffuse alveolar infiltrates. The heart size is normal. There is no evidence of vascular redistribution and Kerley B lines are absent. These are features of ARDS. The patient has an endotracheal tube and a central venous catheter. In this condition there is a pan-endothelial failure resulting in a leakage of fluid from the intravascular space into the alveoli. The common causes of ARDS include septic conditions like severe pneumonia, multiple fractures, massive blood transfusion, near drowning, and pancreatitis.

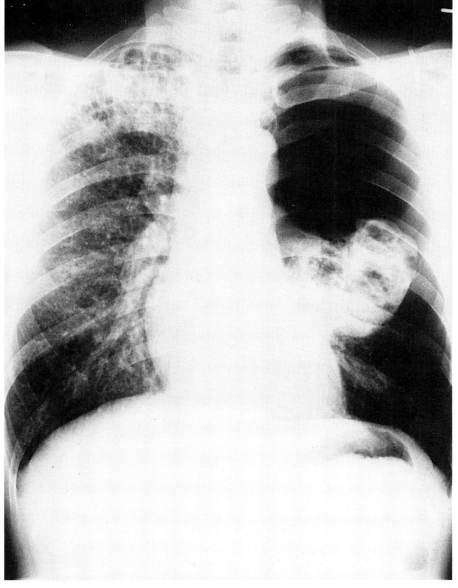

Fig. 86.1

Case 86. This 30-year-old male complained of sudden onset of left-sided chest pain and shortness of breath. Direct questioning also revealed that he had loss of weight and chronic cough of three months' duration. His CXR is shown (Fig. 86.1).

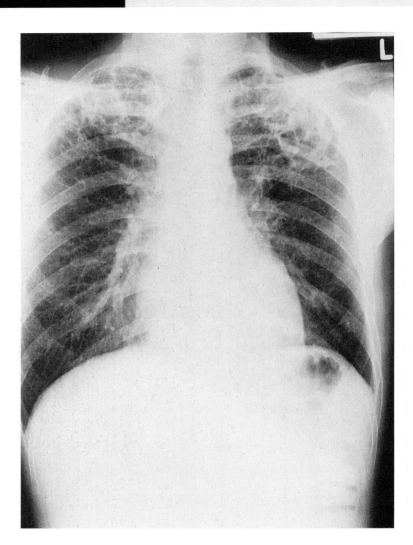

Fig. 86.2

CASE 86 LEFT PNEUMOTHORAX DUE TO PCP

The CXR shows an obvious left pneumothorax. However, the right lung also shows cystic changes in the upper lobe. Following left chest tube insertion, the follow-up CXR (Fig. 86.2) also shows similar changes in the left upper lobe. Subsequent bronchoalveolar lavage showed *Pneumocystis carinii*. Upper lobe cystic form of *Pneumocystis carinii* typically appears in those with HIV who are on prophylactic nebulized pentamidine.

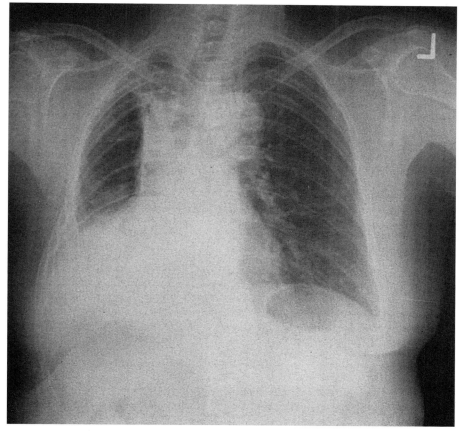

Fig. 87.1

Case 87. This patient gave a history of lung cancer surgery a few years ago followed by radiotherapy. Her CXR is shown (Fig. 87.1).

Fig. 87.2

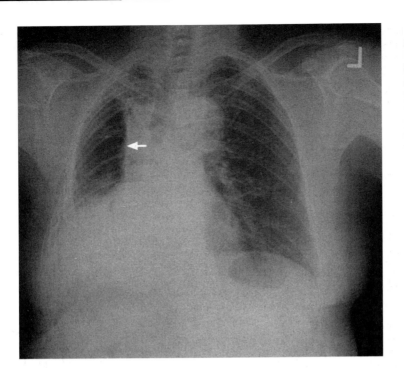

CASE 87   RADIATION FIBROSIS OF RIGHT UPPER LOBE

The CXR shows volume loss in the right hemithorax in keeping with previous right middle and lower lobectomy for cancer. The infiltrates in the upper zone are due to radiation fibrosis and the sharp margin (Fig. 87.2) delineates the limits of lung shielding during the radiotherapy.

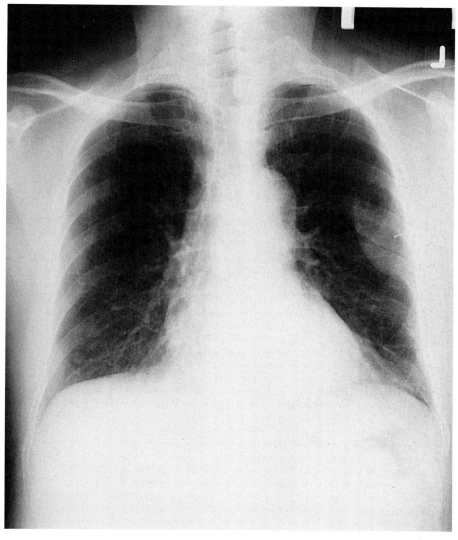

Fig. 88.1

Case 88. This patient is asymptomatic. The CXR is shown (Fig. 88.1).

Fig. 88.2

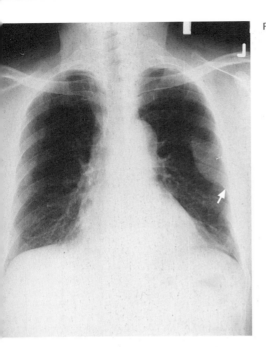

Fig. 88.3

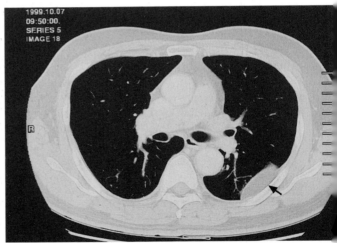

CASE 88 PLEURAL LIPOMA

The CXR shows a left pleural mass. It is usually possible to differentiate this from lung parenchymal masses abutting the pleura which tend to be relatively round with the angle between the mass and the chest wall being acute (unlike here where the angle is obtuse – Fig. 88.2). The CT Thorax (Fig. 88.3) shows that the mass is of fat density confirming the presence of a pleural lipoma, which is benign.

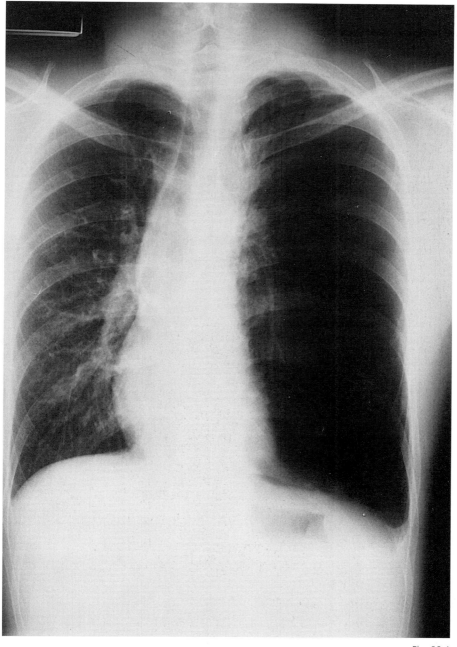

Fig. 89.1

Case 89. This patient is asymptomatic. The CXR is shown (Fig. 89.1).

CASE 89 SWYER-JAMES-MACLEOD'S SYNDROME

The CXR shows typical findings of Swyer-James-MacLeod's Syndrome on the left. The two hallmarks of this condition are unilateral hyperlucency and hypoplastic pulmonary artery. This rare syndrome usually occurs following viral bronchiolitis in childhood. Pneumothorax and bronchiectasis are known complications.

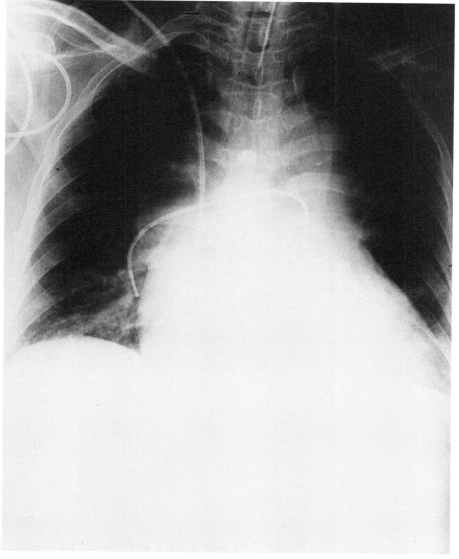

Fig. 90.1

Case 90. This is a routine CXR taken for a patient who was admitted to the Coronary Care Unit (Fig. 90.1) for an acute myocardial infarction.

Fig. 90.2

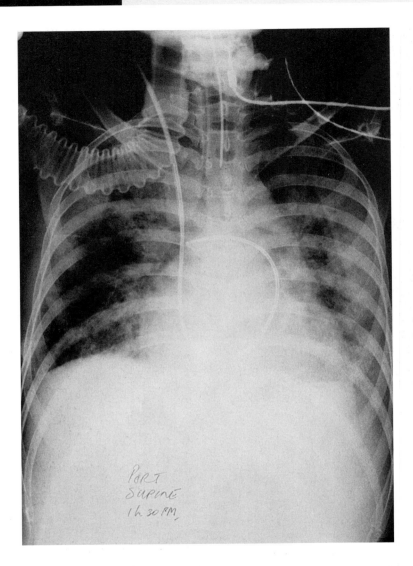

PORT
SUPINE
1 h 30 PM

CASE 90 MALPOSITIONED PULMONARY ARTERY CATHETER (PAC)

The most glaring abnormality is that the PAC has migrated too far distally. The potential complications include knotting of the catheter, perforation of the pulmonary artery, and pulmonary artery aneurysms. The ideal position of the tip of the catheter is just 2–3 cm from the midline (Fig. 90.2) on either side. The first clue to distal migration of the PAC is usually a dampening of the pulmonary artery waveform, which must be monitored on a real-time basis.

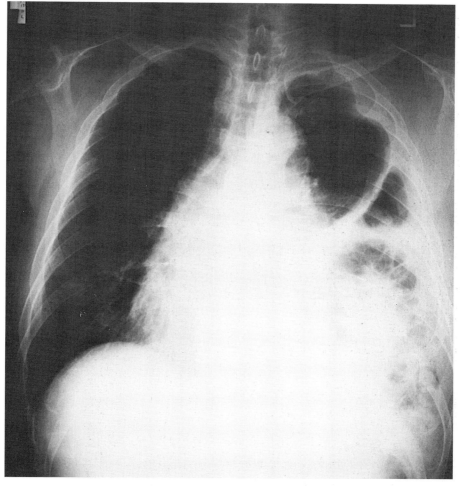

Fig. 91.1

Case 91. This patient gave a history of being involved in a motor vehicle accident resulting in a crush injury to his torso. He is admitted now for vague abdominal pain. His CXR is shown (Fig. 91.1).

Fig. 91.2

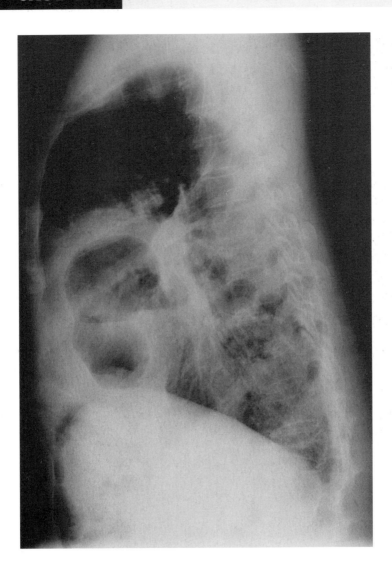

CASE 91 LEFT DIAPHRAGMATIC HERNIA

The CXR shows bowel shadows in the left hemithorax. On the lateral CXR, the bowel shadows seem to be above the left hemidiaphragm (Fig. 91.2). These are features of a left diaphragmatic hernia, which may be congenital or acquired. Barium contrast studies can be used in doubtful cases. Diaphragmatic hernias are typically on the left as the liver is present on the right. Acquired causes are usually traumatic or post-operative. Surgical treatment is usually indicated for traumatic cases.

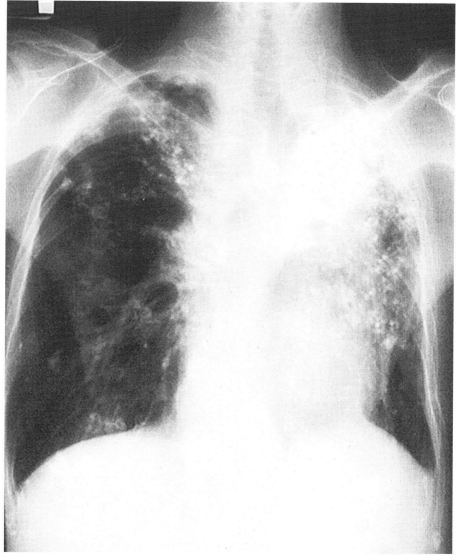

Fig. 92.1

Case 92. This 70-year-old female had past history of pulmonary tuber-culosis for which she was treated twenty years ago. She now presented with streaky hemoptysis of three months' duration. This was her CXR (Fig. 92.1). What is the diagnosis?

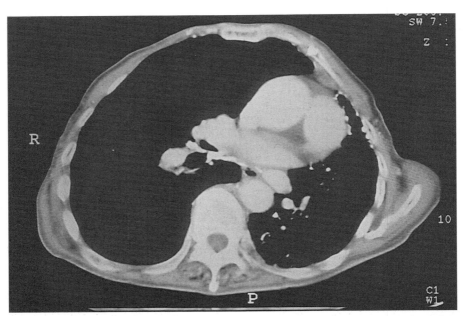

Fig. 92.2

CASE 92 BRONCHOLITHIASIS

The CXR shows severe volume loss and calcification in the left upper lobe consistent with previous tuberculosis. In addition, there is widespread calcification of the mediastinal lymph nodes. In rare instances, a calcified lymph node can erode into the airway causing hemoptysis or obstruction, as shown on CT (Fig 92.2). This condition is called broncholithiasis and can be demonstrated on CT or flexible bronchoscopy. Bronchoscopic removal of the broncholith may be attempted.

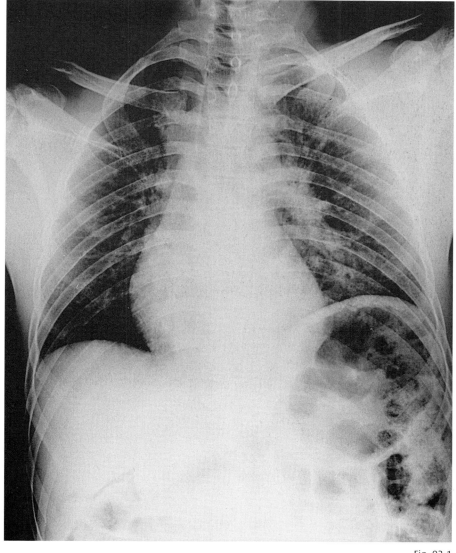

Fig. 93.1

Case 93. This pedestrian was hit by a truck and admitted with severe pain to the left shoulder. His CXR is shown (Fig. 93.1).

Fig. 93.2

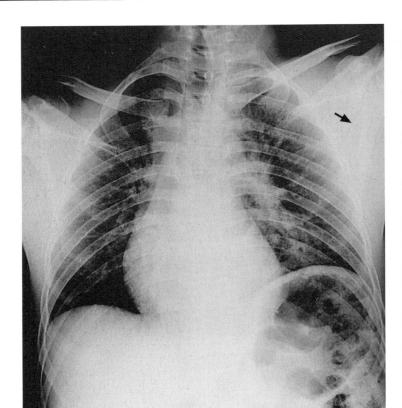

CASE 93 FRACTURED LEFT SCAPULA

The CXR shows an obvious fracture of the body of the left scapula (Fig. 93.2) which usually results from severe trauma. Fracture of the scapula is rare as the overlying muscles provide some protection. The left hemidiaphragm is elevated because of splinting due to pain.

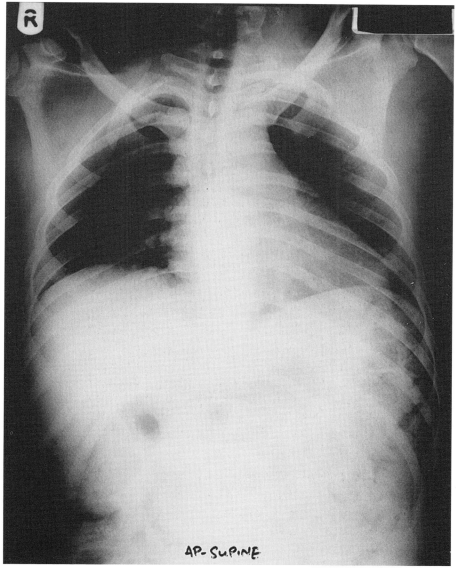

AP- SUPINE

Fig. 94.1

Case 94. This patient was admitted following an accident in which a heavy overhead object fell on him. The CXR is shown (Fig. 94.1).

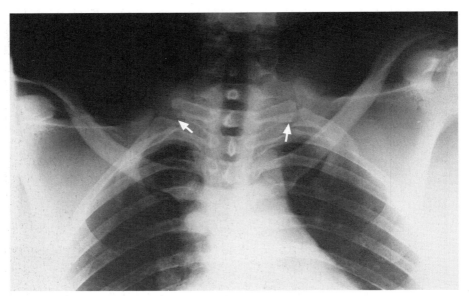

Fig. 94.2

CASE 94 BILATERAL FIRST RIB FRACTURES

The most obvious abnormality is that the first rib margin is discontinuous bilaterally. This is better shown on the apical lordotic view (Fig. 94.2). First rib fractures are usually high-impact injuries as they tend to be protected by the clavicle. They may also be associated with cervical spine and other serious head and neck injuries. The consequence of the fracture is also important as distortion of the anatomy of the blood vessels and nerves may occur.

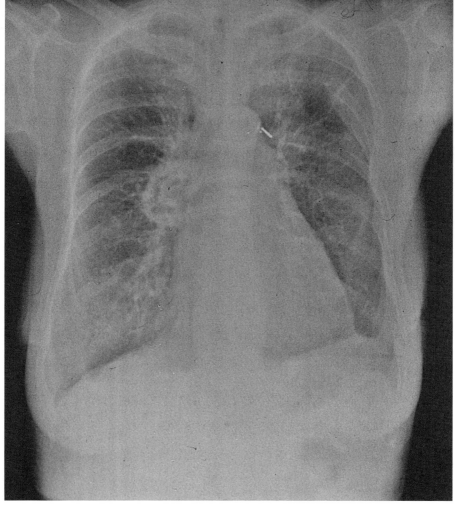

Fig. 95.1

Case 95. This patient presented with cough and fever of one month's duration. She is a known case of COPD with a past history of surgery to the left lung. The CXR is shown (Fig. 95.1).

Fig. 95.2

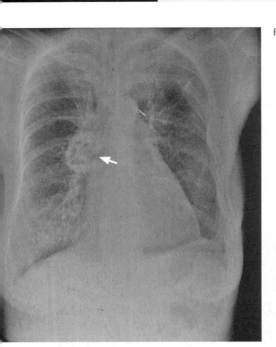

Fig. 95.3

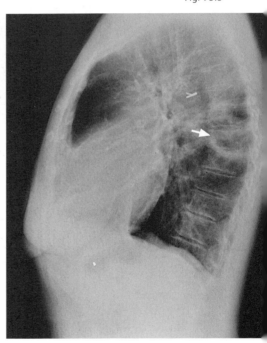

CASE 95   LUNG ABSCESS

See Case 8. The CXR shows a cavitary lesion at the right hilum (Fig. 95.2). Lateral X-ray (Fig. 95.3) shows the lesion to be thin walled and at the apical segment of the right lower lobe. The causes of lung cavities include primary lung cancer (typically Squamous cell), tuberculosis, Klebsiella, *Staphylococcus aureus* (usually multiple), anaerobes, mycetoma, Wegener's granulomatosis, rheumatoid nodule, and pulmonary infarction. Lesions in the upper lobe and apical segment of the lower lobes are typical of pulmonary tuberculosis.

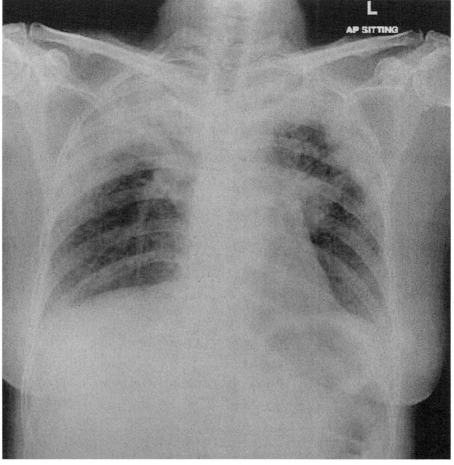

Fig. 96.1

Case 96. This patient presented with a three-month history of coughing and loss of weight. Blood tests showed a markedly elevated peripheral eosinophil count. Her CXR is shown (Fig. 96.1).

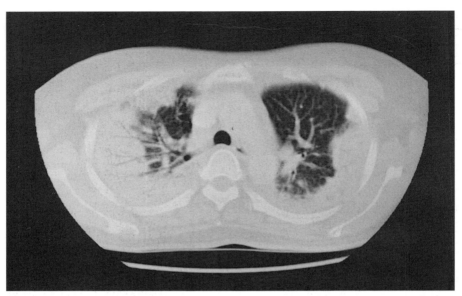

Fig. 96.2

CASE 96   CHRONIC EOSINOPHILIC PNEUMONIA

The CXR shows bilateral upper lobe peripheral consolidation typical of idiopathic chronic eosinophilic pneumonia. The CXR pattern is sometimes described as a radiographic negative of pulmonary edema as illustrated by the CT (Fig. 96.2). Other causes of pulmonary infiltrates and eosinophilia include tropic pulmonary eosinophilia (due to microfilaria), Loeffler's syndrome (due to parasites like ascaris), Churg-Strauss syndrome, allergic bronchopulmonary aspergillosis (ABPA), hypereosinophilic syndrome, drugs, and malignancy. Idiopathic chronic eosinophilic pneumonia is very responsive to steroids.

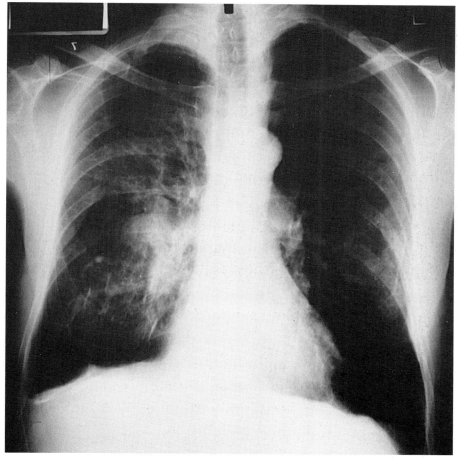

Fig. 97.1

Case 97. This elderly patient was asymptomatic. He used to work on-board a ship as an electrician for many years. The CXR is shown (Fig. 97.1).

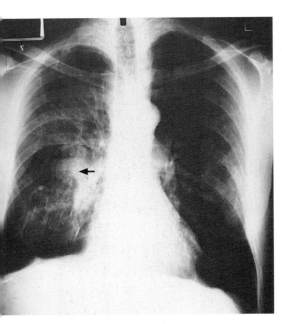

Fig. 97.2

Fig. 97.3

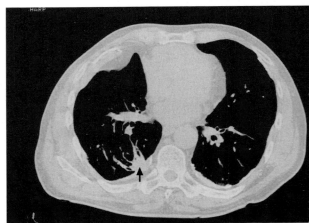

CASE 97 ROUND ATELECTASIS DUE TO ASBESTOS EXPOSURE

The CXR shows bilateral pleural calcification suggestive of previous asbestos exposure. In addition, there seems to be a mass in the right middle zone (Fig. 97.2). CT
Thorax shows the mass to be abutting the posterior pleural surface. The mass (Fig.
97.3) appears to have a comet tail – curvilinear bronchovascular shadows leading
towards it. The diagnostic criteria for round atelectasis are: (1) mass abutting
pleura, (2) comet tail of bronchovascular markings leading towards it, (3) inflam-
matory process involving the pleura (in this case previous asbestos exposure).

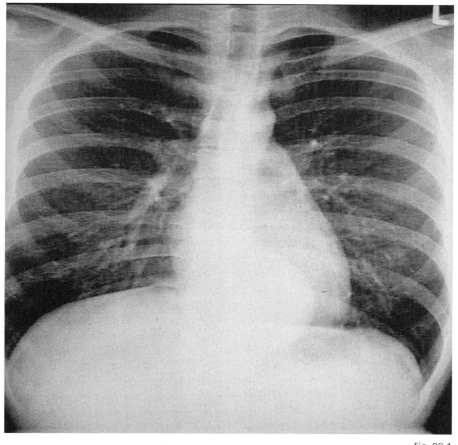

Fig. 98.1

Case 98. This patient was totally asymptomatic. The CXR is shown (Fig. 98.1).

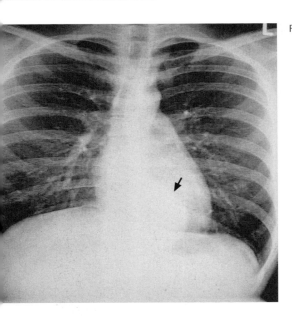

Fig. 98.2

Fig. 98.3

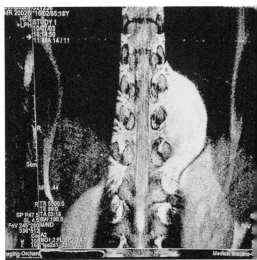

CASE 98 POSTERIOR MEDIASTINAL MASS

The CXR shows a well-circumscribed shadow (behind the heart) abutting the vertebral body and aortic arch (Fig. 98.2) localizing it to the posterior mediastinum. The commonest cause of posterior mediastinal masses are of neurogenic origin, e.g. neurofibroma, neurilemma, ganglioneuroma. Other less common causes include meningocele, extramedullary hemopoiesis, esophageal cysts, and aortic aneurysms. MRI (Fig. 98.3) confirms the presence of the neurogenic tumor.

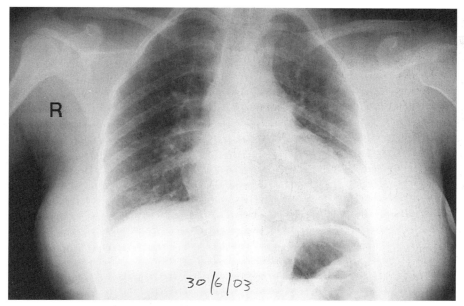

30/6/03

Fig. 99.1

Case 99. This 21-year-old female was cyanotic. Her CXR is shown (Fig. 99.1).

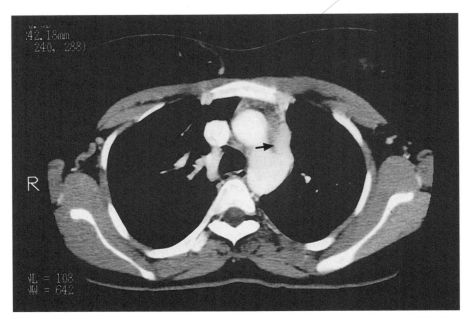

Fig. 99.2

CASE 99 EISENMENGER'S SYNDROME DUE TO
PATENT DUCTUS ARTERIOSUS

The CXR shows that the heart is enlarged and the apex is lifted off the left hemi-diaphragm, typical of right ventricular hypertrophy. The pulmonary arteries are also enlarged. These features are in keeping with those due to severe pulmonary hypertension. The presence of cyanosis in this setting indicates a reversal of blood flow at the site of shunting resulting in flow from the right to the left circulation (Eisenmenger's syndrome). In this patient, the CT (Fig. 99.2) confirmed a patent ductus arteriosus (PDA) as the likely cause.

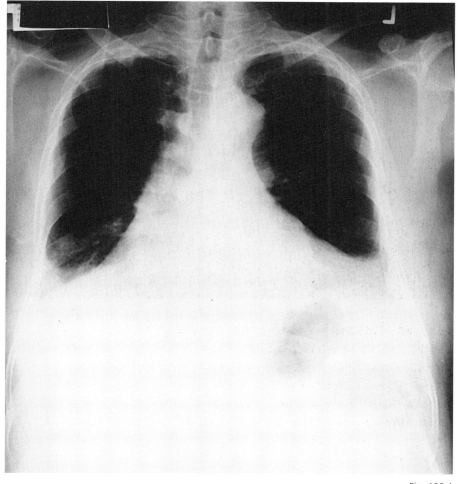

Fig. 100.1

Case 100. This patient is asymptomatic. The CXR is shown (Fig. 100.1).

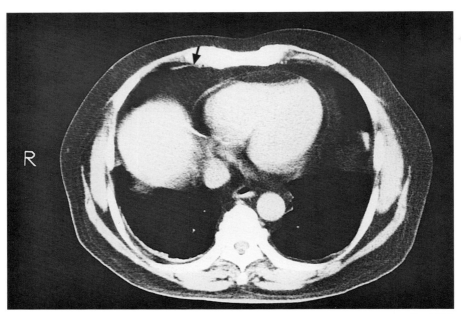

Fig. 100.2

CASE 100 MEDIASTINAL LIPOMATOSIS

The CXR shows widening of the cardiac silhouette with blunting of both cardio-phrenic angles. There also seems to be a blurring of the left heart margin. The CT Thorax (Fig. 100.2) shows widespread fat deposits on both sides of the mediastinum. This condition, called mediastinal lipomatosis, is usually associated with severe obesity, exogenous steroids, and high ethanol intake.

INDEX